ORGANIC FARMING
FOR SUSTAINABLE AGRICULTURE

Prof. Pravin Chandra Trivedi (b. 3 March, 1953), Ph.D., Post-Doct. (U.S.A.), F.L.S. (London), F.B.S., F.P.S.I., F.B.R.S., F.M.A., F.E.S., F.N.R.S., F.I.A.T., Professor, Department of Botany, has a brilliant academic career standing first in all examinations.

He worked as Post-Doctoral Fellow at the Department of Plant Pathology, North Carolina State University, Raleigh, U.S.A. in 1983-84. Prof. Trivedi has 36 years of teaching and research experience in the University of Rajasthan, Jaipur. He has published more than 250 research papers and review articles in journals of repute. He has edited and written many books and guided 40 Ph.D. students. He has been the Principal Investigator of 21 major research projects funded by national organisations. He is an elected Fellow of nine reputed academic bodies. Dr. Trivedi visited fifteen countries for academic purpose and delivered invited talks in many international conferences.

ORGANIC FARMING FOR SUSTAINABLE AGRICULTURE

Editor

Prof. P.C. Trivedi

Ph.D., Post Doct. (U.S.A.), F.L.S. (London), F.B.S., F.P.S.I., F.B.R.S., F.M.A., F.E.S., F.N.R.S., F.I.A.T.
Professor of Botany
University of Rajasthan
Jaipur-302 055, India

Aavishkar Publishers, Distributors

Jaipur 302 003 (Raj.) India

Published by
Prem C. Bakliwal for
Aavishkar Publishers, Distributors
807, Vyas Building, Chaura Rasta
Jaipur 302 003 (Raj.) India
Phone : 0141-2578159
e-mail : aavishkarbooks@hotmail.com

2023

ISBN 978-81-7910-339-5

Printed at
Trident Enterprises (UP)
Noida

Dedicated to

My Parents

Late Shri D.S. Trivedi

and

Smt. Kamla Trivedi

"The Inspirations Behind My Life"

Preface

Organic Farming is a production system of crops which avoids the use of synthetic and chemical inputs like fertilizers, pesticide, growth regulators and live stock feed additives. Organic Farming systems depend mainly on crop rotations, crop residues, animal manures, legumes, green manures, off-farm organic residues, mineral bearing rocks and biological pest control to maintain soil productivity and supply important nutrient like N, P, K. Indiscriminate use of chemical fertilizers for decades has resulted in the decline of organic matter content of the soils to less than one per cent. The use of pesticides has led to pest resurgence and difficulty in controlling weeds. Indiscriminate use of artificial fertilizers and pesticide has caused severe water and environmental pollution. Organic farming ensures the production of palatable food free from chemical residues.

Organic farming means adopting techniques that maintain soil fertility indefinitely, that utilizes as far as possible, only renewable resources, that do not grossly pollute the environment and that fosters life energy within the soil and throughout the cycles of all the involved food grains. The term organic farming is not directly related to the types of inputs used, but refers to the concepts of the farm as an organism, in which all the component parts—the soil minerals, organic matter microorganism, insects, plants, animals and humans interact to create a coherent whole.

Organic farming means farming in the spirit of organic relationship. The concept of organic farming is dubious to the farmers of developing countries. In India, it implies the use of organic manures and no use of synthetic fertilizers and chemical pesticides. It is the method of farming which primarily aims for cultivating land and raising crops in such a way as to keep the soil alive. This method of farming help :

- In maintaining environment health by reducing the level of pollution.
- It reduces human and animal hazard by reducing the level of residue of the product.
- It helps in keeping agricultural production at a higher level and makes it sustainable.
- It reduces the cost of agricultural production and also improves the soil health.

It ensures optimum utilization of resources for short-term benefit. Organic farming is a system of farming based on integral relationship. So one should know the relationship between soil, water, plants, soil microflora and overall relationship between plant and animal kingdom of which man is the apex animal. The totality of these relationship is the backbone of Organic Farming.

Sustainable agriculture is agriculture that is ecologically sound, economically viable and it conserves the resources such as soil, ground and surface water, minerals, petroleum and biodiversity. It also provides the needs of present generation without compromising the ability of future generation to meet their needs. To sustain the agriculture productivity, it becomes imperative to take care of soil health and environment as such. Sustainable crop production is only possible when the natural resources based on which the production activity depends should not be eroded or harmed in any manner. Organic farming is the wise answer to the aforesaid problems.

Only 30 per cent of India's total cultivable area is covered with fertilizer where irrigation facilities are available. The remaining 70 per cent of the arable land, which is mainly rainfed area, has not been using any fertilizer. It is estimated that around 600 to 700 million tones of agricultural waste is available in the country every year but most of it is not properly used. There are several alternatives for vermi-compost, biofertilizers etc. Technology have been developed to produce large quantities of vermicompost. There are specific biofertilizers for cereals, millets, pulses & oilseeds. However, such biofertilizers and biopesticides have not become very popular in India for two reasons. The first is the lack of marketing and distribution network. Retailers are typically not interested in selling bio inputs because their demand is low, supply is erratic and farmer are ignorant about bio-inputs. The second reason is the presence of chemical fertilizers and pesticides, with their heavy advertisements, public support and higher margin for retailers. All these problems need to be addressed for popularizing organic farming in India.

The present book "Organic Farming for Sustainable Agriculture" covers the application of Organic Farming in the present scenario and suggesting the ways and means of driving the maximum benefits by using biofertilizers for sustainable agriculture. Major chapters included in this book are on—Global development of organic agriculture; Biofertilizers—An important component in organic farming; Use of microalgae & Cyanobacteria as biofertilizers; Organic farming for sustainable agriculture development; Vermicompost : Role in Organic Farming and Biodynamic Farming : Holistic approach in agriculture. A detailed chapter on organic farming and Biotechnology provide sufficient information about use of organic farming in the era of biotechnology and how biotechnology is changing the agriculture sector. This book will be useful to agriculture scientist, farmers, botanist, industrialist, social workers and persons interested in organic farming for sustainable agriculture.

I am thankful to all the learned contributors for their co-operation in compiling useful information on various facets of organic farming. Heartfelt thanks go to these contributors, each of whom has endeavoured to present up-to-date information of his area of expertise and has given willingly of very valuable time and knowledge. The opinions and text contained herein are those of the authors and I have tried to honour their ideas in the original shape. I am sure the information given in this book will prove helpful to those interested in Organic Farming for the benefit of human life and human welfare. Let us live in harmony with our mother because we are her children.

I record my sincere gratitude to my wife Kusum, Rohit-Neha (Son & daughter in law), Priyanka-Kumar Gaurav (Daughter & Son in law) for their consistent support in my academic and scientific endeavours and encouragement and love during the preparation of this book.

I cannot forget little angel Kyna for giving me complete refreshment wherever I feel tired by her naughty activities & pleasant smile.

I appreciate the all round co-operation and support of Mr. Akshay Jain of Aavishkar Publishers, Distributors for printing & publishing this book with patience, care and interest.

Jaipur **Prof. P.C. Trivedi**

Contents

Contributors

Behera, Mohima Prasad

Department of Agronomy, College of Agriculture, Orissa University of Agriculture and Technology , Bhubaneswar-751 003

Deokar, C.D.

Department of Plant Pathology and Agricultural Microbiology, Mahatma Phule Krishi Vidyapeeth, Rahuri-413 722

Khan, J.B.

Algal Biotechnology Lab, Department of Botany, University of Rajasthan, Jaipur-302 055

Kumar, Manoj

Algal Biotechnology Lab, Department of Botany, University of Rajasthan, Jaipur-302 055

Mukherjee, Dhiman

Regional Research Station, Uttar Banga Krishi Viswavidyalaya, Kalimpong (Darjeeling)-734 301

Pareek, Aparna

Department of Botany, University of Rajasthan, Jaipur-302 055

Roul, Pravat Kumar

Department of Agronomy, College of Agriculture, Orissa University of Agriculture and Technology, Bhubaneswar-751 003

Sahoo, Jogeswar

Department of Agronomy, College of Agriculture, Orissa University of Agriculture and Technology, Bhubaneswar-751 003

Sharma, Parul

Department of Botany, University of Rajasthan, Jaipur-302 055

Singh, G.P.

Algal Biotechnology Lab, Department of Botany, University of Rajasthan, Jaipur-302 055

Sonawane, R.B.

Department of Plant Pathology and Agricultural Microbiology, Mahatma Phule Krishi Vidyapeeth, Rahuri-413 722

Trivedi, P.C.

Department of Botany, University of Rajasthan, Jaipur-302 055

1

Global Developments of Organic Agriculture : Challenges and Prospects

APARNA PAREEK AND P.C. TRIVEDI

Organic farming is a type of agriculture that benefits from the recycling and use of natural products. Extensive use of dried foliage and kitchen compost not only reduces the investment overheads, but also ensures the growth of crops that are devoid of synthetic interference. The technique is characterized by the use of green manure, biological pest control methods and special cultivation techniques to maintain soil productivity. Limiting the use of synthetic fertilizers or completely doing away with them reduces the risk of exposure to ailments that arise on account of synthetic pesticides, ingrained plant growth regulators and the presence of genetically modified organisms in organic food products.

Today, organic farming caters to a major and preferred industry, worldwide. Organically grown food products have a huge market catered to, by farmlands covering approximately 10% of the total world-farmland cover. The initial effort made by Sir Albert Howard, the Father of Organic Farming, has paid off, metamorphosing into a practice that sustains soil health and ecosystems, by relying on biodiversity, ecological processes and the progress of innovative health sciences.

ADVANTAGES OF ORGANIC FARMING

- The economics of organic farming is characterized by increasing profits via reduced water use, nutrient-contamination by pesticides, reduced soil erosion and carbon emissions and increased biodiversity.

- Organic farming produces the same crop variants as those produced via conventional farming methods, but incurs 50% lower expenditure on fertilizer and energy, and retains 40% more topsoil.
- This type of farming effectively addresses soil management. Even damaged soil, subject to erosion and salinity, are able to feed on micro-nutrients via crop rotation, inter-cropping techniques and the extensive use of green manure.
- Farming the organic way enables farmers to get rid of irksome weeds without the use of any mechanical and chemical applications. Practices such as hand-weeding and soil enhancement with mulch, corn gluten meal, garlic and clove oil, table salt and borax not only get rid of weeds and insects, but also guarantee crop quality.
- The use of green pesticides such as Neem, and compost tea is environmentally friendly and non-toxic. These pesticides help in identifying and removing diseased and dying plants in time and subsequently, increasing crop defense systems.

ORGANIC FARMING DISADVANTAGES

- In 1998, increased risk of *E. coli* infection via consumption of organic food rather than non-organic food was publicized by Dennis Avery of the Hudson Institute.
- A 2008 survey and study conducted by the UN Environmental Program concluded that organic methods of farming result in small yields even in developing areas, compared to conventional farming techniques.
- The Father of the Modern Green Revolution, Norman Borlaug, argues that while organic farming practices are capable of catering to the demands of a very small consumer fraction, the expanding cropland is dramatically destroying world ecosystems.
- Research conducted by the Danish Environmental Protection Agency revealed that organic farms producing potatoes, seed grass and sugar beet are barely able to produce half of the total output churned out from conventional farming practices.

- Organic agriculture is hardly able to address or combat global climate change. Though regenerative organic farming practices are recognized as effective strategies for reducing CO_2 emissions to an extent, the impact is not dramatic.

Nevertheless, despite the various advantages and disadvantages of organic farming, farms where organic methods for cropping are experimented with, display more profitability than conventional farms. Organic agriculture is definitely more sustainable in the long-term, improving soil fertility and terrain drought resistance greatly. These farming practices completely waive off external costs, incurred due to investment in chemical pesticides and nutrient runoff, and a number of health issues that result from agro-chemical residue. The historical perspective of organic farming is given in Table no.1

TABLE 1

Historical Perspective of Organic Farming

Ancient Period	
Oldest practice	10000 years old, dating back to Neolithic age, practiced by ancient civilizations like Mesopotamia, Hwang Ho basin etc.
Ramayana	(All dead things - rotting corpse or stinking garbage returned to earth are transformed into wholesome things that nourish life. Such is the alchemy of mother earth - as interpreted by C. Rajagopalachari).
Mahabharata (5500 BC)	Mention of *Kamadhenu*, the celestial cow and its role on human life and soil fertility.
Kautilya Arthashastra (300 BC)	Mentioned several manures like oil cake, excreta of animals.
Brihad-Sanhita (by Varahmihir)	Described how to choose manures for different crops and the methods of manuring.
Rig Veda (2500-1500 BC)	Mention of organic manure in Rig Veda 1, 161, 10, 2500-1500 BC, is Green Manure

	in Atharva Veda II 8.3, (1000 BC). In Sukra (IV, V, 94, 107-112) it is stated that to cause healthy growth the plant should be nourished by dungs of goat, sheep, cow, water as well as meat. A reference of manure is also made in *Vrksayurveda* by surpala (manuscript, oxford No 324 B, Six, 107-164).
Holy Quran (590 AD)	At least one third of what you take out from soils must be returned to it implying recycling or post-harvest residue.

The various milestones in organic farming are mentioned in Table 2.

TABLE 2

Key Milestones on Organic Farming in Current Period

Sir Albert Howard (1900-1947)	father of modern organic Agriculture, developed organic composting process (mycorrhizal fungi) a Pusa, Samastipur, India and published document" An Agriculture Testament".
Rudolph Stenier (1922)	a German spiritual Philosopher built biodynamic farm in Germany.
J.I. Rodel (1950), USA	Popularized the term sustainable agriculture and method of organic growing.
IFOAM	Establishment of 'International Federation of Organic Agriculture Movement", in 1972
One Straw Revolution	Release of the book by Masanobu Fukoka (1975), an eminent microbiologist in Japan.
EU Regulation	EU Regulations on Organic Food, 1991
Codex	Codex guideline on organic standard, 1999

TABLE 3

Nutrient Use Efficiency in India

Nutrient	Efficiency (%)
Nitrogen	30-50
Phosphorus	15-20
Potassium	70-80
Zinc	2-5
Iron	1-2
Copper	1-2

Organic farming is being practiced in 100 countries of the world. The ill-effects of chemicals used in agriculture have changed the mindset of some consumers of different countries who are now buying organic with high premium for health. Policy makers are also promoting organic farming for restoration of soil health and generation of rural economy apart from making efforts for creating better environment. The global organic area is 26 million hectare roughly along with 61 standards and 364 certification bodies roughly. The world organic market is now 26 billion US$. The organic area in India is 2.5 million hectare including certified forest areas. Non-certified organic area is more than certified organic area. India has developed National Standard under NPOP programme.

The National Centre of Organic Farming under Ministry of Agriculture is promoting organic farming as facilitator across the country and providing various assistance to organic entrepreneurs and farmers.

Organic farming is gaining gradual momentum across the world. Growing awareness of health and environmental issues in agriculture has demanded production of organic food which is emerging as an attractive source of rural income generation. While trends of rising consumer demand for organics are becoming discernible, sustainability in production of crops has become the prime concern in agriculture development.

DEFINITION AND OBJECTIVES OF ORGANIC FARMING

The concept of organic farming is not clear to many concerns. Many people think that traditional agriculture, sustainable agriculture,

Jaivik Krishi etc, are organic farming. Some people are of the idea that the use of organic manures and natural methods of plant protection instead of using synthetic fertilizers/ pesticides are organic farming. But this is not true. The organic farming in real sense envisages a comprehensive management approach to improve the health of underlying productivity of the soil.The system takes local soil fertility as a key to successful production. As a logical consequence, International Federation of Organic Agriculture Movements (IFOAM) stresses and supports the development of self-supporting systems on local and regional levels.

MAIN PRINCIPLES OF ORGANIC FARMING

The main principles of organic farming are the followings:

To work as much as possible within a closed system, and draw upon local resources.

(i) To maintain the long-term fertility of soils.

(ii) To avoid all forms of pollution that may result from agricultural techniques.

(iii) To produce foodstuffs of high nutritional quality and sufficient quantity.

(iv) To reduce the use of fossil energy in agricultural practice to a minimum.

(v) To give livestock conditions of life that confirm to their physiological need.

(vi) To make it possible for agricultural producers to earn a living through their work and develop their potentialities as human being.

The four pillars of organic farming are:

(1) Organic standards

(2) Certification/Regulatory mechanism

(3) Technology packages

(4) Market network

STANDARD AND CERTIFICATION

The most important aspect in modern era of organic farming is certification programmes which consist of standards (rule), inspection (checking whether the rules are implemented) and certification (judgment). Only by this certification programme, organic farming can be distinguished from other methods of sustainable agriculture. These standards define what can be labeled 'certified organic' and sold commercially as such. Infact, certification in organic agriculture generally refers to independent third party certification. Third parts imply it is not done by either the producer (first party) or the buyer (second party). The system includes farm inspector and audit trails (checking of record). Certificate is valid only if it is done by accredited certifying agency. Certification programme vary among countries or regions because of differences in environmental, climatic, social and cultural factors. Needless to say, from a commercial perspective it is not enough that product is produced organically, what is equally important is that it should be certified as such.

TECHNOLOGY PACKAGES

Conventional practices cannot be followed for growing crops organically. It includes land preparation, selection of variety, organic fertilization, biological control of pest- diseases-weed, harvest, storage etc. Some countries have developed package of practice for some selected crops but there is ample scope to refine this package with scientific methods and practices.

MARKET NETWORK

Organic farming has a place where there is a market to accept the produce at a higher price as the growing interest in organic farming practice is due to an expectation of higher premium for organically produced farm commodities. The basic focus of Organic farming should be first to produce a farm products for the home (domestic) market and second for the export market.

GLOBAL SCENARIO OF ORGANIC FARMING

In approximately 100 countries of the world, organic farming is being practiced and the area under organic management is continuously growing.

Area

As per the study (2004) of the Foundation Ecology and Agriculture (known as SOEL), the global organic area is 24 million ha. The major part of this area is located in Australia (about 10 million hectares), Argentina (almost 3million hectares (Table 4)). Australia /Oceania holds 42% of the world's organic land, followed by Latin America

(24.2%) and Europe (23%). In Africa, more than 3,20,000 hectares and 7I ,000 farms are managed organically, representing about 0.04% of agricultural land. The total organic agricultural area in Asia is now about 8, 80,000 hectares, corresponding to 0.07% of the agricultural area. The number of organic farm is more than 61,000.

TABLE 4

Land Area of Major Countries Under Organic Management (SOEL - Survey, 2004)

S. No.	Name of the country	Area under organic (m.ha.)	% of total agricultural area	Number of organic farm
1.	Australia	10.0000	2.20	1380
2.	Argentina	2.9600	1.70	1779
3.	USA	0.9500	0.23	6949
4.	UK	0.7245	4.22	4057
5.	Germany	0.6969	4.10	15,628
6.	South Africa	0.0045	0.05	250
7.	China	0.3012	0.06	2910
8.	Japan	0.0005	0.10	-
9.	India	0.0370	0.03	5147
10.	Pakistan	0.0002	0.08	405
11.	Sri Lanka	0.0015	0.65	3301
	All World	24.0700	1.60	462475

The recent (2005) survey of SOEL indicates that currently more than26 million hectares of farmland are under organic management worldwide. The increase is almost 10% as compared to previous year. The maximum organic area is lying with Australia (11.3 million ha.) followed by Argentina (2.8 million ha)

Standard

Globally there are more than 60 standards which include IFOAM basic standards, CODEX Alimentations Commission guidelines, EU Regulation 2029/91, NOP of USA etc. (Table 5).

TABLE 5

Some Facts on International Standards

1. IFOAM	• Established in 1972 • Headquarter in Germany • Umbrella organization for organic Agriculture Association • Developed international basic standards of organic agriculture • Established IFOAM accreditation programme (1992) to accredit certifying bodies • Set up International Organic Accreditation Service (IOAS) in July 2001
2. CODEX	• Codex Alimentations Commission - a joint FAO/WHO • Intergovernmental body • Established in 1962 • Produced a set of guidelines for organic production
3. EU regulation	• Laid out a basic regulation for European Union's organic standards in Council regulation NO 2092/91 (June 1991) • Regulations give guidelines for the production of organic crops in the European Community.
4. Demeter	• Demeter International is a world wide net work of 19 international certification bodies in Africa, Australia, Europe • Developed guideline for biodynamic preparation.
5. JAS	• A set of guidelines 'Japan Agricultural Standards' for organic production.

Certification

As per Organic Certification Directory, 2003 as published by

Grolink, there are 364 certification bodies across the world (Table 6), but they are unevenly spread. 290 of them are located in European Union, USA, Japan, Canada & Brazil. There is IFOAM accreditation programme, launched in 1992, by initiating International Organic accreditation service (lOAS).

TABLE 6

Certification Bodies and Their Approval Per Region

Region	Total	IFOAM	Japan	ISO 65	EU	USA
Africa	7			1		
Asia	83	4	65	1	1	2
Europe	130	10	9	45	100	28
Latin America/ Caribbean	33	4		10	5	8
North America	101	4	1	14		64
Oceania	10	4	6	3	6	4
Sum	364	26	81	74	112	106

Source : (13)

Market

The current (2004) world organic market is estimated at over US $ 26 billion cultivated on a total area of around 24 million hectares world wide). It will reach 31 billion US$ by 2005. Premiums on most organic products range between 35-100%. Organic products are almost entirely (over 95%) consumed in developed countries. The major producers and importers of organic products are EU, USA and Japan. Categories of major organic products include fresh fruits and vegetables (non tropical and tropical), cereals (wheat, rice, com, maize), coffee, tea, cocoa, spices, herbs, oilseeds, pulses, milk product, honey, meat, edible nut, semi-processed fruits etc.

ORGANIC FARMING IN INDIA

Background : The Indian Agriculture is traditionally organic and farmers were following organic cultivation till the middle of the last century (1950). The Green Revolution was ushered in India during sixty and it has been the cornerstone of India's agricultural achievement, transforming the country from the stage of food deficiency to self-sufficiency. During the period, the production of food grains has increased four folds, from 50.82 met in 1950-51 to 212.05 met on 2003-04. But indiscriminate and excessive use of chemicals during this period has put forth a question mark on sustainability of agriculture in the long run calling attention for sustainable production which will address soil health, human health and environmental health and eco-friendly agriculture..

The year 2000 is very important year for India from organic point of view. The four major happenings were made during the year 2000. These are :

- The Planning Commission constituted (2000) a steering group on agriculture who identified organic farming as National challenge and suggested it should be taken in the form of a project as major thrust area for 10th-plan. The group recommended organic farming in NE Region, rain fed areas and in the areas where the consumption of agro chemicals is low or negligible.
- The National Agricultural Policy (2000) recommended promotion of traditional knowledge of agriculture relating to organic farming and its scientific up gradation.
- The Department of Agriculture and Cooperation (DAC), Ministry of Agriculture constituted (2000) a Taskforce on organic farming under the chairmanship of Shri Kunwar Ji Bhai Yadav and this task force recommended promotion of organic farming.
- The Ministry of Commerce launched the National Organic Programme in April 2000 and Agricultural and Processed Food Products Export Development Authority (APEDA) is implementing the National Programme for Organic Production (NPOP). Under the NPOP, documents like National standards, accreditation criteria for accrediting inspection and certification agencies, Accreditation procedure, inspection and certification procedures have been prepared and approved by National Steering Committee (NSC).

GOVERNMENT POLICY ON PROMOTING ORGANIC FARMING

The policy of Ministry of Agriculture seeks to promote technically sound, economically viable, environmentally non-degrading, and socially acceptable use of natural resources in favor of organic agriculture. The policy seeks to actualize the area and crop potential for strengthening rural economy, promoting value addition, accelerating growth of agro-business and securing a fair standard of living for the farmers and agricultural workers and their families.

AREA APPROACH AND CROP IDENTIFICATION

For promotion of organic farming, identification of potential areas and crop is crucial. The Government strategy to promote organic farming for the crops having market potential like fruits, spices, oilseeds, pulses, vegetables, wheat, cotton, basmati rice, etc. As far as potential areas are concerned, three priority zones have been identified.

Category-I

The top priority areas for promotion of organic farming are the rain fed areas where fertilizer and agro-chemical consumption is already very low.

Category-II

The Category II areas are primarily under rain fed farming with little irrigation support.

Category-III

The last priority areas are those with moderate to heavy use of fertilizers and pesticides, mostly multiple cropped areas.

Apart from promoting organic farming in low consumption area, rain fed area, dry land, hill area etc., some efforts are also being made in high intensive cropping area like Punjab, Haryana etc. for organic farming. But it requires sufficient study to access its acceptability in these areas.

NATIONAL PROJECT ON ORGANIC FARMING

The Department of Agriculture and Cooperation (DAC), Ministry of Agriculture, Government of India has launched a new Central Sector Scheme 'National Project on Organic Farming' with on outlay of Rs. 57.05 crore for production, promotion and market development of organic farming in the country during the 10th Plan on pilot basis with the following components :

The project is operational since October, 2004. Its Headquarters is NCOF, Ghaziabad which has 6 regional centers (RCOFs) at Bangalore, Bhubaneswar, Hissar, Jabalpur, Imphal and Nagpur.

The unique feature of this scheme is to promote group certification by capacity building through service provider where the service provider will help organic farmers by providing training, educating on record management, supervising internal control system, assisting in certification and market access. One service provider will be registered with 1500 farmers working in identical geographical condition with similar type of crop.

OTHER PROJECTS OF MOA IN SUPPORT OF ORGANIC FARMING

Apart from DAC project on organic farming, the projects like 'Technology Mission on Horticulture', Cotton scheme etc. is supporting organic farming. Besides, under Macro Management programme, Government of India is supporting different State schemes on organic agriculture.

FAO-DAC PROJECT ON ORGANIC FARMING

The Ministry of Agriculture has taken up the Technical Cooperation Programme (TCP) of FAO (TCP/IND/3003 - Development of Technical Capacity Base for the Promotion of Organic Agriculture in India) to overcome the knowledge gap by quickly producing basic information tailored to various ecological zones of the country on the current state of knowledge on organic crop production packages, input production and utilization and certification issues including legal and institutional aspects.

ORGANIC FARMING APPROACH BY NAAS

The National Academy of Agricultural Sciences (NAAS) has issued a Policy Paper on Organic Farming, which concludes that while synthetic pesticides can be avoided, complete exclusion of fertilizers may not be advisable under all situations. NAAS recommends that a "holistic approach involving Integrated Nutrient Management (INM), Integrated Pest Management (IPM), enhanced input use efficiency and adoption of region- specific promising cropping systems would be the best organic farming strategy for India." To begin with, the practice of organic farming should be for low volume, high value crops like spices, medicinal plants, fruits and vegetables. NAAS has also emphasized the need for intensive research on soil fertility and plant health management and on issues relating to microbial contamination of food arising from the use of farm yard manures. (Organic Farming: Approaches and Possibilities in the context of Indian Agriculture, Policy Paper 30, NAAS, February 2005).

TABLE 7

Organic Crops of India

Organic Crop	Area	Producers
Tea (Orthodox, CTC, Green)	Darjeeling, Assam, Door, Nilgiris	Plantations
Coffee (Arabica, Robusta)	Western Ghats, Nilgiris	Plantations and individual small farmers
Spices :		
Pepper and cardamom	Western Ghats, Nilgiris	Plantations and individual small farmers
Cloves, nutmeg, mace,	Kerala, Tamil Nadu	Small farmers + farmers' organizations
Garlic, ginger, turmeric	Throughout India	Small farmers + farmers' organizations
Fenugreek, fennel, cumin, coriander	Throughout India	Small farmers + farmers' organizations
Sesame	Gujarat	Contract farmers
Coconuts	Along all coasts	Small farmers
Fruits :		
Apples, pears, peaches,	Himachal Pradesh and Kumaon	Small and medium farmers
Mango, bananas, pineapple, papaya	Maharashtra, Madhya Pradesh	Small and medium farmers
Apricots	Himachal Pradesh and Kumaon	Small and medium farmers
Dry fruits :		
Chestnuts	Jammu and Kashmir	Small farmers, one processor
Pear, mango	Kodaikanal	One processor
Cashew nuts	Pondicherry, Tamil Nadu	Small farmers, few processors

Contd...

Contd...

Organic Crop	Area	Producers
Vegetables :		
Potatoes	Kumaon hills	Small farmers
Cabbage	Kumaon hills	Small farmers
Wheat, maize and sorghum	Haryana, Punjab, Madhya Pradesh, Uttar Pradesh, Maharashtra	Farmers and farmers' groups
Rice :		
Basmati	Haryana	Medium and big farmers
Other varieties	Throughout	Small and medium farmers
Pulses	Throughout	Small farmers
Oilseeds	Madhya Pradesh	Medium and big farmers
Coarse grains	North-east, Gujarat, Orissa	Small farmers
Honey	Forest areas	Individual collectors
Cotton	Gujarat, Madhya Pradesh	Farmers' organizations

THE ROLE OF AGRICULTURE AND RURAL DEVELOPMENT IN INDIA

Contribution of agriculture to the national economy

As in most developing countries, agriculture still is a cornerstone of the Indian economy. It provides 29 per cent of gross domestic product (GDP), as against industry at 26 per cent and services at 45 per cent (Tata Services 2001). The net value added for agriculture and allied activities amounts to 4,112 billion rupees, or US$ 92 billion (Tata Services 2001). Out of 329 million hectares, the net cropped area is 142.5 million hectares (Ministry of Agriculture 2000).

National agricultural policy

This agriculture policy tries to integrate all the various aspects.

It mentions that over 200 million Indian farmers and farm workers have been the backbone of India's agriculture and that, despite having obtained national food security, the well-being of the farming community continues to be a matter of grave concern. It further aims at :

- A growth rate in excess of 4 per cent a year
- Growth, based on efficient use of resources and conservation of soil, water and biodiversity
- Growth with equity, in other words, widespread across regions and farmers
- Growth that is demand driven and caters to domestic markets and maximizes benefits from exports
- Growth that is technologically, environmentally and economically sustainable

The approach to rural development and land reforms will focus on the following areas :

- Consolidation of holdings on the pattern of the north-western states
- Redistribution of ceiling surplus lands and waste lands among the landless farmers and unemployed youth with initial start-up capital
- Tenancy reforms to recognize the rights of the tenants and sharecroppers
- Development of lease markets for increasing the size of the holdings, by making legal provisions for giving private lands on lease for cultivation and agribusiness
- Updating and improvement of land records, computerization and issue of pass-books to the farmers and
- Recognition of women's rights in land.

The policy has the form of an approach paper and just mentions each and every thinkable aspect related to agriculture. A number of committees have been set up to operationally the various aspects.

However, it was learnt from the Ministry of Agriculture that deliberations are on going and no definite results have been reached as yet.

Rural infrastructure

Electrification of villages has been a priority for the last 25 years and most villages, which can be reached by road, are now connected. This, of course, does not mean uninterrupted supply. In most cases a 4- 8 hours daily supply is common. In 1995/96 road length, both surfaced as well as un-surfaced, was reported to be 3,320,000 kilometers. However, in remote areas there are still villages unconnected by road. In 1998/99 there were 137,800 post offices in the rural areas, while the number of inhabited villages is 627,000 (Tata Services 2001).

Rural employment and labor situation

In 2001, the total population was estimated at well over 1 billion, of which the rural population constitutes around 70 per cent, or 700 million. The total rural working population is around 350 million: 123 million (35 per cent) are cultivators, 84 million (24 per cent) are agricultural laborers and the remaining 41 per cent are other workers (Ministry of Agriculture 2000). Other main areas of work are in the remaining part of the primary sector, such as forestry, fishing, plantations and allied activities and furthermore in household industries as weavers, potters, etc. (Tata Services 2001). It should, however, be realized that many workers migrate to towns and cities during the lean agricultural season and work as coolies, cycle rickshaws and in construction.

Male migration leads to increasing female responsibility for the economic maintenance of families. The female agricultural laborers are worse off than women in cultivating households in terms of employment. While only 10 per cent of cultivating households wanted employment to add to family resources, more than 50 per cent among female agricultural workers were seeking work. Seasonal fluctuations in employment affect women much more than men.

Women in agriculture and rural areas

In the rural areas, the work participation rate as a percentage to the total population in 1991 was 53 per cent male workers and 27 per cent female workers (Tata Services 2001). Women's contribution to farming is insufficiently recognized and agriculture policy is still dominated by the false view that 'farmers are men', women are only housewives. Their work in agriculture tends not to be recorded as 'work' or as 'production' because it falls outside the so-called 'production boundary'.

There is a close relationship between rural poverty and the high incidence of female agriculture labor. The majority of landless women laborers in agriculture have poor literacy, irregular employment and a heavy work burden. Thus a growing imbalance exists between women's access to land, labor, capital, services and facilities on the one hand and the demands of production on the other.

In an agricultural economy, land is the main asset. Every state has set ceiling limits on land holdings, yet the skewed distribution of land ownership shows how around 9 per cent of rural households control almost 44 per cent of the total cultivated land.

The Indian livestock population is also steadily increasing. In 1992 the number of stock animals amounted to 471 million, or about one animal for two people. Cattle, buffaloes and goats constitute the majority (Ministry of Agriculture 2000). The predominant farming and production systems include:

- Plantations: Restricted to tea, coffee, rubber, spices, bananas
- Farming by farmers with help of agricultural laborers
- Share-cropping: Land is leased to tenants by the owner for cultivation and agricultural produce is shared by both. Only in certain areas has this traditional system been in use for generations
- Contract farming: Predominantly in sugar cane, rice and cotton. It operates with credit and buyback- guarantees to the farmer. It is assuming increasing importance especially in organic farming because of the requirement of a steady supply. Training and certification is arranged by the contractor. Mainly adopted by private companies
- Forest collection: Mainly by tribal people. Collection includes honey and herbs for medicinal purposes

Environmental problems and trends

A World Bank study estimated that the major environmental costs for India are 4.5 per cent of GDP. Water pollution, soil degradation and urban air pollution have the highest share, followed by rangeland degradation, deforestation and tourism. What has not been accounted for, however, is the loss of biodiversity, the loss of soil productivity, and health costs arising out of pollution caused by industrial wastes and use of pesticides.

In 1983, about 70 per cent of all pesticides used in India were banned or severely restricted in Western countries and identified by the World Health Organization (WHO) as hazardous. This figure came down to 54 per cent in 1996 and is expected to gradually decrease. However, it is a fact that T-BHC, T- Endosulphan, Dieldrin, Aldrin and even DDT are still widely used in agriculture (Agarwal 1999). Numerous studies carried out by Indians in India and abroad indicate the high impact of pesticides on the immune system. The studies were held in rural and urban areas and include cotton field workers and factory workers exposed to pesticides and mothers.

Socio-economic background

The population below the poverty line in rural areas has steadily decreased from 37.3 per cent, or 244 million, in 1993/94 to a projected 18.6 per cent, or 138 million, in 2001/02. However, there is a huge difference in the respective states. The states with 40 per cent and above in rural areas are Bihar (58 per cent), Orissa (50 per cent), Assam, Manipur, Nagaland, Sikkim and Tripura (45 per cent), Uttar Pradesh (42 per cent) and West Bengal (41 per cent) (Tata Services 2001). In these eastern states, women and children bear a disproportionate burden of poverty. In regions of the highest agricultural growth (north-west), there is the least incidence of poverty, but discrimination against the female population persists. The southern states are rated somewhere in-between.

India has witnessed phenomenal agricultural growth that has wiped out earlier food grain deficits and has led to the accumulation of food stocks. Today India has about 60.4 million tons of food grains in its stock against the buffer norms of 15.8 million tons. Yet this has had no significant impact on reducing the incidence of absolute poverty in rural India because rising output has been accompanied by rising prices owing to price support policies and extensive unemployment, which has reduced the purchasing power of the poor.

Child labor is common in all farming communities and in household industries. However, it has decreased in factories. Statistics are difficult to obtain.

ORGANIC FARMING IN THE NATIONAL CONTEXT

Institutional framework

Historical development of organic agriculture and its key factors

Traditional practices : Traditional agriculture in India dates back to the Neolithic Age of 7,500-6,500 BC. The farmers of Ancient India are known to have evolved nature friendly farming systems and practices such as mixed farming, mixed cropping and crop rotation. The balance of cosmic forces, health and fertility were the main characteristics. Hindu philosophy regards the earth as a living being. She is considered the source of all plants, especially crops, and when cultivated or explored, provides all necessities of life not only for human beings, but also for all other forms of life, right from the smallest living cell to the largest animals. Farmers' knowledge of plant life was highly advanced.

Classical Indian plant science, Vrikshayurveda in the form of Sanskrit hymns, is a corpus of rich textual knowledge. It encompasses areas such as the collection, selection and storage of seeds, germination, sowing, various techniques of plant propagation, grafting, nursing and irrigation, testing and classification of soil and selection of soil suitable to various plants/types of plants, managing, pest and disease management/ preventive and primitive care to build up disease resistance and to cultivate healthy plants. Favorable and unfavorable meteorological conditions were taken care of. Plants were used as indicators of weather, water, minerals etc. This knowledge system is even today present with millions of Indian farmers as its practitioners. Furthermore, it is propagated in many forms, such as folk songs, rituals, proverbs and riddles. Organic agriculture practices make use of these indigenous knowledge systems and try to integrate them in the modern organic agricultural practices, thus making changes easier and more effective.

Historical evidence indicates high yields in India comparable to today's highest levels which was a result of the careful husbanding of soil and well adapted seeds and crop varieties. In the eighteenth century, Thomas Barnard, a British engineer, who conducted a survey of 800 villages near Madras, reported that if the average yield of wetland rice was 3.6 tons per hectare, in 130 villages it was 8.2 tons per hectare while in many villages the yields surpassed 10 tons a hectare. The present day average is 3.1 tons with extensive use of chemical fertilizers and pesticides. Against insects and disease, the main weapon of these farmers has been genetic diversity. India once had 30,000 varieties of rice. These varieties were not used at random, but were delicately fitted into their appropriate ecological niches.

In the vision of this tradition, there is no distinction between the sacred and the profane: everything is sacred. Farmers paid a great deal of attention to agriculture, livestock, rain, harvest and the art of composting. The wisdom gained and practices adopted by these farmers were passed down through generations and became ingrained in the cultural outlook of the society. Even today the belief system, the myths, rituals and religious festivals of the Indians encompass these principles of soil, plant and animal health. Villagers worship Earth in the form of cow dung, symbolizing the fertility of moist soil and mysterious powers of growth. The practices of seed selection, soil fertility and biodiversity are interwoven in the belief system, the rituals and dietary system of the communities.

In the past five decades, the traditional knowledge and organic principles were eroded because of the influx of modern conventional agriculture. However, this knowledge has been sustained by many Indian communities throughout the millennia and has gained renewed importance recently for present agriculture, especially organic agriculture. Organic farming practices still are a part of the living tradition of most of the Indian communities in the Tribal and dry land areas. With organic production and trade fast increasing globally, there is a growing interest in organic agriculture in the country. Traditional practices in India see the earth as a living being and there is still reluctance to exploit the earth for short gains. Traditional agricultural practices obviously can be improved and organic agriculture is the closest to the farmer's traditional customs, practices and beliefs. Organic agriculture can make the existing practices more conscious. It is, however, also true that conventional extension services have not reached these dry land areas. Chemical fertilizers and pesticides are not easily available and even if they are, small and marginal farmers would hardly be in a position to buy them.

In 1983, the first training centre in organic agriculture was set up in Pondicherry under a project called agriculture, man and ecology (AME), implemented by Educational Training Consultants, Leusden, and the Netherlands and financed by the Government of the Netherlands. In October 1984, the Association for the Propagation of Indigenous Genetic Resources organized the first conference on organic farming in Wardha. In 1992, the Rajasthan College of Agriculture organized a national seminar on natural farming. Since then, numerous farmers turned organic and important networks, such as ARISE (Agricultural Renewal in India for a Sustainable

Environment), were established. In 1993, a directory of individuals and organizations involved in sustainable agriculture in India, called *Green Farming* was produced (Centre for Science and Environment 1993). In 1994, a register of 365 Indian organizations was published (ICEIA/ETC India 1994). Not all organizations were involved in organic agriculture, but all were at least related.

In a few agricultural universities (Hisar, Bangalore, Coimbatore and Madurai), organic agriculture is promoted by individual faculty members. Many conferences followed. Biodynamic agriculture got a boost, mainly in South India and Darjeeling (tea estates). The export of organic coffee, tea, spices and later Basmati rice increased. The central government set up a special cell for the export of certified organic products under APEDA of the Ministry of Commerce and Industries. In March 2000, the same Ministry launched the national programme of organic production. In October 2000, the Ministry of Agriculture constituted the task force on organic agriculture.

In June 2001, under the national programme of organic production, a set of four volumes, concerning accreditation regulations, accreditation criteria, accreditation procedure and application forms were published. On 12 June 2001, by Public Notice No.19 (RE-2001/1997-2002) the government introduced regulations for the export of organic produce. It was stated that an agricultural product would be allowed to be exported as an "organic" product only if it was produced, processed or packed under a valid organic certificate issued by a certifying agency duly accredited by one of the following accreditation agencies: APEDA, the Coffee Board, the Tea Board and the Spices Board. On 16 June 2001, an advertisement appeared in the Economic Times of India calling inspection and certification bodies for accreditation by the government.

Regulatory framework (standards, inspection, certification)

Previously exports were certified by foreign certification bodies such as the Institute for Market Ökologie (IMO) (Switzerland), SKAL (the Netherlands), Ecocert (France) and Naturland (Germany). The Department of Commerce under the Ministry of Commerce and Industry is the nodal agency that has laid down the national programme for organic production and declared the standards for organic products. It will accord recognition to the standards of other nations and seek recognition from other national bodies for Indian standards. Standards are modeled after the IFOAM (International Federation of Organic Agriculture Movement) Basic Standards. Their scope includes production, processing, handling, labeling and animal husbandry. "India Organic" was established as the logo.

As the institutional arrangements for implementing the programme, a Steering Committee has been set up as an apex advisory body. The Steering Committee consists of representatives of the Ministries of Agriculture, Food Processing Industries, Forests and Environment, Science and Technology, Rural Development and Commerce, and Trade and Exports. The Committee's functions are:

- To lay down the national accreditation policy
- To set out the general philosophy and principles of organic production
- To prescribe the eligibility criteria for inspection agencies
- To authorize accreditation agencies toward the national organic logo to those found eligible to carry the official logo

The accreditation criteria are the same as those of IFOAM. The accreditation agencies comprise the Tea Board, Coffee Board, Spices Board and APEDA. Each of the three commodity boards prescribes the practices for the production of their respective crops, while APEDA has prepared them for rice, sugar cane and passion fruit. The accreditation agencies are assisted by a common evaluation agency in screening inspection and certification bodies and checking their compliance to norms of organic production.

CHALLENGES OF DOMESTIC CERTIFICATION

The regulation aimed to help domestic organic producers to overcome international trade barriers and reach a "level playing field". Yet Indian regulation does not simplify the requirements for Indian exporters, nor does it reduce the costs. Agricultural products from India can be marketed as organic within the European Union (EU) under two exporting options for third countries: 'Equivalence Granted' and 'Imports Granted'. Under the first option organic products can be exported if their production and inspection systems are considered as equivalent to those of the EU. That means India has to be recognized as having equivalent standards to those in the EU as defined in EU Regulation 2092/91. Recently India has requested its inclusion to the third countries approved list and will have to wait for the EU resolution on its application.

Until the approval is granted and India is added to a granted list, despite following national norms for organic agriculture, producers will have to comply with EU production norms also. Secondly, since Indian local certifiers are not recognized by the EU, certification has to be issued by an EU approved certification body (by law already approved by the control authority in the EU importing country). Further, exporters in third countries are not allowed to apply directly for import authorization. So Indian exporters have to depend on importers in individual EU member states to obtain special import permits from their respective EU control authorities. Moreover, import permits are issued for a defined period either for specific products or product groups from a given EU country. This operates as technical barriers to trade and increases transaction costs, continuing to hamper international trade.

The Japan national standards for organic products became official on 1 April, 2000. The full implementation of United States national standards is expected soon. Thus a similar compliance problem will be faced in export to the United States and Japan. Experience has shown that it can take several years for the negotiations to be worked out. In addition, one has to go through this process with every country to which export is intended.

Key actors

Central government : A number of key central government actors have already been noted above. In addition, the Ministry of Agriculture should be mentioned. It took a long time for this important Ministry to take an active part in the promotion of organic agriculture. In October 2000, it was decided to set up a task force on organic agriculture, separate from the Steering Committee functioning under the Ministry of Commerce. The main objectives of this task force are:

- To advise the Ministry of Agriculture on all aspects of organic agriculture
- To promote organic agriculture in general

The task force was to submit its report in June 2001, but the date was postponed to August 2001. However, at the time of finalizing this study, the report is still awaited. It is hoped that the influence of the task force on the implementation of the National Agriculture Policy will be substantial.

Organic agriculture began as an ideology, but can it meet today's needs?

There is a widespread belief that low-yielding organic agricultural systems are friendlier to the environment and more sustainable than high-yielding farming systems. The current aims of organic systems—maintenance of soil fertility, avoidance of pollution, use of crop rotation, animal-welfare concerns and wider environmental aspects—would be hard to quarrel with.

Does organic farming benefit biodiversity?

The intensification and expansion of modern agriculture is amongst the greatest current threats to worldwide biodiversity. Over the last quarter of the 20th century, dramatic declines in both range and abundance of many species associated with farmland have been reported in Europe, leading to growing concern over the sustainability of current intensive farming practices. Purportedly 'sustainable' farming systems such as organic farming are now seen by many as a potential solution to this continued loss of biodiversity and receive substantial support in the form of subsidy payments through EU and national government legislation.

ORGANIC FARMING AND THE SUSTAINABILITY OF AGRICULTURAL SYSTEMS

The desire for a sustainable agriculture is universal, yet agreement on how to progress towards it remains elusive. The extent to which the concept of sustainable agriculture has any operational meaning is discussed. Sustainability is considered in relation to organic farming — a sector growing rapidly in many countries. The role of regulation and the use of synthetic agrochemicals, the desired degree of self reliance of agricultural systems, and the scale of production and trade in agricultural goods are all considered in the context of this discussion of sustainability.

Organic farming has proven beneficial for many but the yield of organic farming has not been substantial since its beginning. Many farmers have turned to organic farming only on account of the subsidies given by the government to promote organic farming. There also have been instances wherein farmers have opted for organic farming on account of reduced production cost as compared to traditional farming. Low productivity, higher time required yielding and requirement of specialized skills for implementation have been some of the disadvantages of organic farming. However, organic farming has been significantly contributing towards providing quality food and prevention of soil erosion which have nowadays become serious concerns in many regions of the world.

REFERENCES

Agarwal, Anil, 1999. *State of India's Environment 5: The Citizens' Fifth Report,* part I and II, (New Delhi, Centre for Science and Environment).

APEDA, 2001. Questionnaire for collection of information for formulating assistance schemes for organic farming.

APEDA, 2001. Proceedings of the National Awareness Programme on Organic Farming, Bhubaneshwar, India.

APEDA, 2001. "The export market for organic products from India" internal note, APEDA, New Delhi.

Bhattacharyya, P. 2004. Organic Food Production in India - Status. Strategy and Scope. Agribios (India, Jodhpur), 01 - 182.

Dudani, A.T. 1999. *Alternatives to Pesticides in Tropical Countries* (New Delhi, Vigyan Prasar).

Export - Import Bank of India. Export of Organic Products from India: Prospects and Challenges (Occasional Paper No.97), 1- 166.

Fertiliser Statistics, FAI, 2003-04.

Government of India, Ministry of Commerce and Industry, 2000. *National Programme for Organic Production,* containing the standards for organic products, New Delhi.

Government of India , 2001. *National Programme for Organic Production: Accreditation Regulations* (New Delhi).

Government of India, 2001. *National Programme for Organic Production: Procedures for Accreditation* (New Delhi).

Government of India, 2001. *National Programme for Organic Production: Accreditation Criteria for Inspection & Certification Agencies* (New Delhi).

Government of India, 2001. *National Programme for Organic Production: Formats and Application Forms* (New Delhi).

Government of India, 2001. Public Notice No. 19 (RE-2001)1997-2002, Ministry of Commerce and Industry, Department of Commerce, New Delhi.

Howard, Sir Albert, 1940. *An Agricultural Testament,* (Goa, India, republished by The Other India Press).

Lampkin, N. 1990. Organic Farming, Farming Press Book, IPSWICH, UK.

Sharma, P.D. and Singh Mohan. 2004. Problems and Prospects of Organic Farming. Bulletin of Indian Society of Soil Science, No.22, 14-41.

❑❑❑

2

Biofertilizer : An Important Component in Organic Farming

C.D. DEOKAR AND R.B. SONAWANE

Nowadays there is an increasing consciousness about conservation of environment as well as health hazards associated with agro chemicals and consumers presence to safe and hazards free food. This has led to the growing interest in alternate forms of agriculture in the world organic agriculture is one among the broad spectrum of production methods that are supportive of the environment. The demand for organic food is steadily increasing both in developed and developing countries with an annual average growth rate of 20–25 % organic agriculture without about is one of the fastest growing sectors of agricultural production. However, there are certain issues that should be clarified before we go for a large scale conversion to organic agriculture. The most important issues are can organic farming produce enough food for every body? Is it possible to meet the nutrient requirement of crops entirely from organic sources? Are there any significant environmental benefits that occur from organic farming? Is the food produced by organic farming superior in quality? Is it economically feasible?

WHAT IS ORGANIC FARMING

Organic farming is a farm of agriculture which avoids of largely excludes the use of synthetic fertilizers and pesticides, plant growth regulators and live stock feed additives. As far as possible organic farmers rely on crop rotation, crop residues, animal manures and mechanical cultivation to maintain to soil fertility and productivity to supply plant nutrients and to control weeds, pest and diseases.

According to International Organic Farming Organization "The role of organic agriculture, whether in farming processing distribution of consumption is to sustain and enhance the health of ecosystems and micro organisms from the smallest in the soil to a human beings.

HISTORY

Organic agriculture is the oldest form of agriculture on earth. Farming without the use of petroleum based chemical fertilizers and pesticides was the sole option for fares until post world war II. The war brought with it technologies that were useful for agricultural production for example, ammonium nitrate used for ammunitions during world war II evolved in to ammonium nitrate fertilizer, organophospate natural gas production led to the development of powerful insecticides. These technical advance since world war II have resulted in significant economic benefits as well as environmental and social determents. Organic agriculture seeks to utilize those advances that consistently yield benefits (new varieties of crops, precision agricultural technologies, more efficient machinery) while discarding those methods that have led to negative impacts on society and the environment, such as pesticide pollution and insect pest resistance . Instead using synthetic fertilizers and pesticides, organic farmers utilize crop rotations, cover crops and natural base products to maintain or enhance soil fertility. These farmers rely on biological, cultural and physical methods to limit pest expansion and increase populations of beneficial insects on their farm. Because genetically modified organisms constitute synthetic inputs and pose unknown risks among such as herbicide resistance, seeds of plant and product ingredients like GM lecithium are disallowed in organic agriculture.

The world organic farming was first introduced by Lord Northbound. The term is used in his book. "Look to the Land" in which he had put forth suggestions for a wholesome approach to ecologically balanced farming. However, the practice of organic farming has been prevalent from the time agriculture was first conceived. It was with the advent of the "green revolution" that the natural way of growing crops started becoming unpopular and economically less feasible, more crop yield meant higher profit of the farmer and better utilization of the land. At times when only one crop was grown per season or year the farmers who used fertilizers started growing to crops during the same time period.

Organic food farming continue in small mostly family run farms of kitchen gardens where people grew food for there own requirements. The produce was sold in farmers market. Large scale organic farming was began by farmers and scientists as a mark of protest to agricultural industrialization, Now organic food is widely available and has become very popular, with sowing sales.

POTENTIAL AREAS FOR ORGANIC FARMING IN INDIA

About 65% of India's cropped area is not irrigated and it can be safely assumed that high input demanding crops are not grown on these lands. The fertilizer use only on lands is always less anyway as chemical fertilizers required sufficient water. To respond pesticide use in these lands would also be less as the economic of these hardy of not so profitable crops will not permit extensive inputs. These areas are atleast relatively organie of perhaps even "organic by default". While neither of these terms necessarily denotes a healthy forms and recommended agriculture system if would atleast imply a non chemical form that can be converted very easily to an organic one providing excellent yield and without the necessity and efforts of a lengthy conversion period.

TECHNIQUES, PRACTICES AND SPECIALIZED FARM OF ORGANIC FARMING

Techniques and practices of organic farming will change from farmer to farmer and area to area depending on local ecosystem. However, some of the common techniques adopted and popularized are biodynamics, biological / natural disease , pest and weed control , composting, cover cropping, crop rotation, diversity on the farm, do nothing farming, effective microorganism (EM) , green manuring and green leaf manuring , homa farming, indegenous seeds, intercropping, live fences, , microbial biofertilizers, mulching, multi cropping, multi purpose trees, polyculture, reduce tillage, soil and water conservation, vermi-composting, Plant products and specialized organic farming techniques.

Agroecosystem

Ecosystem refers to a biological community of interacting organisms and their physical environment. It represents the totally of relationships among organisms and their ambient surrounding environment. All ecosystems are driven by solar energy and are fixed as organic carbon by green plants. **Agroecosystem** refers to a model for working within an agriculture system with all its inputs and outputs. The pattern of energy and material flow between plants, livestock and humans are simple, consistent and similar in most agricultural system. Agroecosystems have relatively simple cycles. **Agroforestry** or Farm forestry is the cultivation of forest trees on a farm, alongwith conventional crops, such as herbaceous plants. Economic gains are derived from such crops. Agro-silviculture, agro-horticulture, agro-hort-siliviculture, homestead agroforestry, etc are examples of agroforestry systems.

Long lasting challenges in soil microbiology

The global necessity to increase agricultural production from a steadily decreasing and degrading land resource base has placed considerable strain on the fragile agro ecosystem. Current strategies to maintain and improve agricultural productivity via high input practices place considerable emphasis on "failsafe" techniques for each component of the production sequence with little consideration to the integration of these components in a holistic , systems approach. While the use of mineral fertilizers is consider the quickest and surest way of boosting crop production , their cost and other constrains deter farmers from using them in recommended quantities. In recent years, concepts of integrated plant nutrient management (IPNS) have been developed , which emphasize maintaining and increasing soil fertility by optimizing all possible sources (organic and inorganic) of plant nutrients required for crop growth and quality. This is done in an integrated manner appropriate to each cropping system and farming situation. Improvement in agricultural sustainability requires optimal use and management of soil fertility and soil physical properties, both of which rely on soil biological processes and soil biodiversity. An understanding of microbial diversity perspectives in agricultural context, is important and useful to arrive at majors that can act as indicators of soil quality and plant productivity. In this context, the long lasting challenges in soil microbiology of development of effective methods to know the types of micro organisms present in soil and to determine functions which the microbes for perform *in situ.*

Rhizosphere

The presence of soil around the root surface defines the

Fig. : Rhizosphere soil

rhizosphere, while the intimate interactions between the plant and soil biota within the soil habitat characterize the rhizosphere. Root exudates are the substrate or fuel for the intense microbial activity within the rhizosphere. Thus, it is the quantity and quality of the exudates and condition of the soil habitat will determine the colonization potential of the rhizosphere. Each plant species leaks a unique carbon and nitrogen signature of carbohydrates, amino acids and organic acids that determines the primary colonizers of the microbial community. The composition of the exudates also affects the availability of nutrients in the immediate vicinity of the roots influencing the establishment and growth of the plant. An increase or decrease in the nutritional status of the plant can further after the quality of the exudates affecting the microbial diversity and populations in the rhizosphere. Root biomass and architecture, fertilizer type, tillage and cropping history also affect root growth and patterns of exudation and the populations and diversity or rhizosphere colonizing organisms. Together, these factors determine whether rhizosphere interactions and processes will have a positive, neutral or beneficial affect on plant growth. Microbial inoculants placed with or in the immediate proximity of the seed must be able to successfully establish and multiply along the growing root, and integrate into the rhizosphere community or alter and/or overwhelm the community to give the desired growth effect. The diversity of crops in a rotation, perhaps even the sequence of crops in a rotation not to mention the cultivar will affect the efficacy of an inoculant regardless of the crop and soil management techniques. It is suggested that the greatest success will be achieved in developing inoculants that contain cool temperature competitive strains, and a diversity of species that more tailored to a particular crop species or type. Technical considerations aside, one size fits no one. There is also a need to work with plant breeders to develop appropriate inoculants, that work in synergy with the crop traits, particularly rooting.

TABLE 1

Microorganisms Present in Fertile Soil

S.No.	Microorganisms	No./ g. of soil
1	Bacteria	1.50 Crore
2	Actinomycetes	7.00 lakh
3	Fungi	4.00 lakh
4	BGA	50,000
5	protozoa	30,000
6	Urea solubilizers	8500
7	Denitrifying bacteria	400
8	Pectin solubilizers	70000
9	Butyric acid Bacteria (anaerobic)	5000
10	Protein solublizers (anaerobic)	22000
11	Cellulose Decompsers	350
12	Nitrogen fixer (aerobic)	1885
13	Nitrogen fixer (anaerobic)	700
14	Nitrifying bacteria	1701

The leading strategies for restoring quality and productivity of degraded soils and plant rhizosphere are 1) enhancing soil organic matter pool, 2) creating positive carbon and other nutrient budget 3) strengthing nutrient cycling process and 4) improving soil biology. Altering rhizosphere zone for positive biological and chemical balance can be the most promising technology for sustaining the soil health and crop productivity.

The importance of micro organisms in biological nutrients mobilization and plant growth promotion are well known and various micro organisms singly or in combinations are being exploited as biofertilizers. Biofertilization now is a mature biotechnology. The journey started with nitrogen fixing rhizobium now encompasses long list of micro organisms with most important being *Azotobacter, Azospirillum, Acetobacter, Rhizobium, Beijerinckia, Dextria,* blue green algae, *Azolla* and Phosphate solubilizers such as *Bacillus polymyxa, Bacillus megaterium, Bacillus subtilis, Pseudomonas strata, Aspergillus awamori, Penicillium digitatum* etc. Recently another group of micro organism known as plant growth promoting rhizobacteria (PGPR) such as *P. flueorescens* and Phosphate mobilizing mycorrhizal fungi such as *Glomus* spp. have also attracted lot of attention. So far commercial biofertilizers are mono strain bacteria/ fungal cultures. Large numbers of co-inoculation studies have proved that use of multiple microorganism give mush higher benefit compare to single micro organisms.

ROLE OF MICROORGANISM IN ORGANIC FARMING

Soils provide the widest variation of habitats for micro-organisms and, in turn, contain the most varied array of micro-organisms. Microbial modes of life have changed through geological time, and microbial metabolism has accounted for most changes that have occurred. Primitive forms of life were probably simple and unicellular (if cellular at all) and existed under conditions quite different from those of present-day life.

Forms of elements are changed by microbial metabolism. Some important examples are found in cycles of carbon, oxygen, nitrogen and sulfur. Some micro-organisms produce organic carbon compounds from carbon dioxide, and others break down organic carbon to carbon dioxide. The process of changing organic to inorganic substances has been termed **mineralization.** Some micro-organisms change ammonia to NO_2, and others convert NO_2 to NO_3. On the other hand, some utilize NO_3 molecules for electron acceptors and reduce it. Both oxidation and reduction of sulfur also occur. Nitrogen and sulfur both occur in amino acids. Some bacteria fix atmospheric nitrogen directly, but other nitrogen fixers live symbiotically in roots of plants. Micro-organisms account not only for the great quantity of metabolic changes but also for many qualitative changes that are carried out nowhere else in nature.

TABLE 2

Statewise Status of Biofertiliser Production (tonnes)

Name of the State	Biofertiliser production during the year 2005-06							
	Azotobacter	Azospirillium	Acetobacter	Rhizobium	PSB	Total BF	Others*	Grand total
Andhra Pradesh	3.35	19.61	N.A.	34.18	2189.29	2246.43	15.04	2261.47
Assam	N.A.	N.A.	N.A.	N.A.	106.20	106.20	7250.56	7356.76
Bihar	9.00	8.00	N.A.	5.50	18.50	41.00	N.A.	41.00
Gujarat	294.50	119.97	80.19	126.18	750.76	1371.60	N.A.	1371.60
Delhi	0.49	0.06	N.A.	0.18	0.50	1.23	1.09	2.32
Haryana	6.85	N.A.	0.19	6.46	9.98	23.48	N.A.	23.48
Himachal Pradesh	4.29	N.A.	N.A.	5.30	N.A.	93.59	9.59	9.59
Jharkhand	1.5	0.98	N.A.	1.00	5.52	9.00	200	209.0
Karnataka	3.978	4.61	N.A.	13.91	45.34	67.84	0.08	67.92
Kerala	N.A.	4.89	N.A.	0.64	2.81	8.34	8.34	8.34
Madhya Pradesh	136.95	0.51	N.A.	252.17	433.44	823.07	140.57	963.64
Maharashtra	1064.82	37.36	0.85	152.45	843.49	2098.96	394.04	2493.00

Contd...

Contd...

Name of the State	Biofertiliser production during the year 2005-06							
	Azoto-bacter	Azospir-illium	Aceto-bacter	Rhizo-bium	PSB	Total BF	Others*	Grand total
Nagaland	3.39	2.17	N.A.	3.30	8.17	17.03	N.A.	17.03
Orissa	7.98	7.00	N.A.	24.99	26.00	65.97	200	265.97
Punjab	N.A.	N.A.	N.A.	2.27	N.A.	2.27	N.A.	2.27
Rajasthan	42.78	N.A.	83.40	99.57	204.84	430.59	6.18	436.77
Pondicherry	N.A.	1.98	N.A.	3.77	2.03	7.78	1.96	9.74
Tamilnadu	10.87	929.16	0.06	264.96	1002.52	2207.57	77.99	2285.56
Uttar Pradesh	168.21	0.08	15.82	50.55	251.64	486.30	N.A.	486.30
West Bengal	35.857	36.561	N.A.	38.1	84.09	194.60	1285.51	1480.12
Total	**1794.81**	**1172.941**	**180.51**	**1085.48**	**5985.12**	**10218.85**	**9573.02**	**19791.88**

Source : Biofertilizer News letter

Organisms requiring oxygen for respiration are called aerobes, eg. *Rhizobium, Azotobacter* and *Azospirillum*. In aerobic respiration, food stubs (usually carbohydrates), are fully oxidized to carbon di- oxide and water, releasing chemically energy, as in the reaction :

$$C_6H_{12}O_6 + 6O_2 -------- 6\ CO_2 + 6\ H_2O + \text{energy}$$

ROLE OF NITROGEN FIXING BIOFERTILIZERS IN ORGANIC FARMING

Biofertilizers are the microbial preparations containing live cells of bacteria / cyanobacteria / fungi which efficiently fix atmospheric N, solublilize insoluble phosphate in soil or decompose cellulotic organic wastes. The efficient strain of the requisite microbial agent is grown on specific nutrient medium on large scale aseptically, mixed in a suitable carrier and the preparation so formulated is sold as biofertilizer. Many of the microbial inoculants all over the world are based on solid peat formulations. Biofertilizers are also called as bio-inoculants, microbial cultures, bacterial inoculants or bacterial fertilizers. Biofertilizers are grouped in three categories depending on kind of beneficial activity carried out by the microbial agents used in their production. These include : (1) 'N' fixing biofertilizers, (2) Phosphate solubilizing biofertilizers and (3) Decomposing cultures. Here, detail information on nitrogen fixing biofertilizers is given.

Nitrogen fixing biofertilizers

These include the preparations of bacteria / cyanobacteria which convert atmospheric gaseous nitrogen (N_2) to ammonia (NH_3) and make it available to the plants. The organisms, which fix atmospheric nitrogen, are called as diazotrophs. Diazotrophs belong to prokaryotic group and none of the members from eucaryotic group is able to fix atmospheric N_2. The nitrogen-fixing organisms include species of *Rhizobium, Azotobacter, Azospirillum* and blue green algae (cyanobacteria). These are used in commercial production of biofertilizers. The list of N_2 fixers may be extended to include those diazotrophs which are not being used in commercial production. These are *Beijerinckia indica, Derxia gummosa, Clostridium pasteurianum, Klebsiella pneumoniae, Frankia alni, Rhodospirillum rubrum, R.. capsulata, Chlorobium limicola, Chromatium minus, Desulfovibrio* sp. and *Methanobacterium* sp. Recently *Acetobacter diazotrophicalis* is reported to be a potential N fixer in sugarcane.

Azotobacter biofertilizer

Azotobacter is a bacteria capable of fixing atmospheric nitrogen. It is one of the dominant non symbiotic nitrogen fixing bacteria, encountered in neutral to alkaline soil conditions. The ovule shapes azotobacter is present in large numbers in rhizosphere. It does not attach itself to roots because azotobacter synthesizes its nutritional requirement from organic matter, it is called heterotrophic. As it derives its energy from oxidation of organic substances, it is called chemoheterotrophic. They are mono, di and polysacchrides, as well as several other organic compounds as carbon source. They form cysts to overcome adverse condition which are characterized by the presence of poly B- hydroxyl butyric acid. As the *Azotobacter chroococcum* colony gets older, it accumulates polysaccharide or gum, together with a black melanin pigment which is another characteristic feature of the *Azotobacter* spp. The organism grow when the pH is near neutral, but pH below 6.0 hampers the process of nitrogen fixation.

The free living nitrogen fixing bacteria are classified into aerobic, anaerobic and facultative anaerobic types depending on the growth and survival of organisms either in the prescence or absence of oxygen. The aerobic bacteria fixes nitrogen which is an anaerobic process and requires some arrangement to exclude the oxygen from the fixation site. It has the ability to synthesis the auxins, vitamins, growth promoting substances and antifungal antibiotic. Some important growth promoting substance are Indole acetic acid, giberellins and b vitamins. Another important characteristics of azotobacter is that it excretes ammonia in the rhizospher in the presence of root exudates which helps in the nutrient uptake by the plant. Plant root exudates contain certain amino acids, sugars, organic acids and vitamins, which serves as energy sources for *Azotobacter.*

Azotobacter is Gram -ve, non - spore forming bacterium, which fixes atmospheric nitrogen asymbiotically in the rhizosphere of cereals, vegetables and fruit crops. The predominant species *viz., Azotobacter chroococcum* is widely used in carrier based preparations for biofertilizer production. The other species of *Azotobacter* include *A. beijerinkii, A. paspali.* and *A. vinelandii, Azotobacter paspali* in only exception which acts as associative dinitrogen fixer in Paspalum grass. The beneficial effects of *Azotobacter* are not only due to its ability to fix nitrogen but also due to its ability to secrete growth-promoting substances and its capacity to produce antifungal antibiotics, which improves plant stand in inoculated fields by inhibiting root pathogens. There is an increase in yield of *Azotobacter* inoculated crops by 10 to 15 % and saving of N fertilizers by 15 to 20 per cent.

Azospirillum Biofertilizer

The beneficial effect of inoculation with *Azospirillum brasilense* on crop productivity have been widely described, but extensive use in typical agricultural field environments is scarcely documented. The nitrogen fixing rhizobacterium *Azospirillum* lives in close association with plant roots, where it exerts if beneficial effects on the plant growth and yield of many crops of agronomic importance.

Azospirillum is a ubiquitous soil bacterium capable of colonizing effectively in the roots of a wide variety of plants forming apparently associative symbiosis. The bacterium is Gram –ve, vibrio in shape, motile, contains Poly-B-Hydroxybutyrate granules. When grown in nitrogen free medium it behaves as microaerophillic and when supplied with fixed nitrogen it grows as an aerobic organism. *Azospirillum* sp. increases mineral and water uptake of the plants. It associates with, and also helps root development and vegetative cover. The species of *Azospirillum* so far reported are *A. basilense, A. lipoferum, A. amazonense* and *A. seropedicae. Azospirillum* inoculation consistently increased the yield of sorghum, pearl millet, sugarcane, sunflower, mustard, finger millet and vegetables up to 17 to 18 % with saving of 20 to 25 kg N/ha.

Rhizobium biofertilizer

Bacteria present in the rhizosphere of soil are called **rhizobacteria**. They first infect root hairs and subsequently enter the nodule cells to produce methaemoglobin. The later is an oxygen carrying pigment also called leghaemoglobin that makes environment anaerobic, which is necessary for nitrogen fixation. The plant roots supply the nutrients necessary for the bacteria. Rhizobia respire aerobically, and derive food (carbon source) and energy from organic matter which makes them **chemoheterotrophic**. In their symbiotic association with leguminous plants. Rhizobia enter the root hairs, multiply in roots, and form nodules on the roots. Legumes provide the carbon source in monosacchride and di sacchrides) for the bacteria, which in turn fix the atmospheric nitrogen for plants. The nitrogen fixing capacity of rhizobia makes it a very important biofertilizer, although all its strains do not fix nitrogen to the same extent.

The bacteria rhizobia belongs to the genus *Rhizobium* and have many strains. *Rhizobium* is a rod shaped bacterium motile, non-spore former, Gram -ve in staining reaction and capable of fixing atmospheric nitrogen by forming a symbiotic relationship with legume host. The bacterium invades the specific legume crop roots and induces the formation of specialized structures called as nodules. Bacteria multiply inside nodule and bacteriods (the forms of *Rhizobium* inside nodule) convert atmospheric nitrogen to ammonia, which is utilized in the formation of amino acids by legume host.

Taxonomical position of *Rhizobium* according to eighth edition of Bergey's Manual of Systemic Bacteriology is shown under family Rhizobiaceae under Part VII. The family Rhizobiaceae includes genera *viz., Rhizobium* and *Bradyrhizobium*. Genus *Rhizobium* has three species *viz.,* (1) *Rhizobium leguminosarum* with three biovars as *R. leguminosarum* biovar *trifolii* for clover group, *R. leguminosarum* biovar *phaseoli* for bean groups and *R. leguminosarum* biovar *viceae* for pea group, (2) *R. meliloti* for alfa-alfa group and (3) *R. upini* for lupinae group and *Rhizobium* spp. for cowpea group.

The genus *Bradyrhizobium* includes the species *viz., Bradyrhizobium japonicum* for soybean. It is a slow growing, nitrogen fixing strain of bacteria that inoculates soybean to form root nodules for atmospheric nitrogen fixation. The Bradyrhizobium sp. are slow growing strain of nitrogen fixing bacteria inoculating lotus, vigna and *cicers* types of legumes,

TABLE 3

Nitrogen Fixation in Pulses

S. No.	Crop	N (kg/ha/ season)	S. No.	Crop	N (kg/ha/ season)
1.	Alfa-alfa	120	9.	Cluster - bean	56
2.	Pegionpea	140	10.	Methi	32
3.	Sunhemp	90	11.	Soybean	90
4.	Dhencha	69	12.	Groundnut	64
5.	Cow pea	90	13.	Gram	65
6.	Udid	39	14.	Lentil	33
7.	Mung	35	15.	Beans	54
8.	Peas	59			

Blue green algal biofertilizers

Blue green algae, the older name for cyanobacteria, is a heterogenous group of photosynthetic nitrogen fixing organisms containing chlorophyll a, and used as biofertilizer for wet land rice. The colour of algae is because of the preference of the photosynthetic pigment phycocyanin. Besides their ability to fix atmospheric nitrogen, the blue green algae produce some useful substances like auxins, vitamins of B_{12} group, Indole acetic acid and giberellic acid .

The free living cynobacteria, majority of which are heterocystous and nitrogen fixing contribute on an average 20 to 30 kg N/ ha. Synthesis and excretion of organic/ growth promoting substances by the cyanobacteria are also on record. During the last 2-3 decades, a great volume of work has been accomplished on the various important fundamental and applied aspects of cynobacterial biofertilizer which include strain identification, isolation, purification and culture laboratory analysis of their N_2 fixing activity and related physiology, biochemistry and energistic. The identification of structure and regulation of nitrogen fixing (*nif*) genes and nitrogenous enzymes.

The blue green algae are prokaryotic, single celled or filamentous, nitrogen fixing photoautotrophic organisms. They are also called as Cyanobacteria. They have the ability to fix both CO_2 and N_2 using light energy. Nitrogen fixing unicellular blue green algaes are the *Gleocapsa, Xanococcus, Aphanothece, Pleurocapsa* and *Myxosarcina*. Filamentous non - heterocystous blue green algal species belong to the genera *Plectonema, Oscillatoria, Trichodesmium, Pseudoanabaena, Microcoleus* and *Scizothrix*. The blue green algae, which are filamentous and possess heterocysts, include the species of anabaena, *Aulosira, Nostoc, Scytonema, Calothrix, Tolypothrix, Fischerella, Cylindrospermum, Anabaenopsis* and *Stignoems*.

Mass production of blue green algae, for small to large scale biofertilizer production, involves four methods which utilize soil based algae inoculants and are a) pit method, b) trough method c) field method and d) nursery-cum-algae production method. Blue green algae are also commercially available in the form of dried algal flakes. They may contribute 25-30 kg of nitrogen per hectare per cropping season depending on ecological conditions. Ten days after the transplantation of rice the dried algal flakes are applied to the soil @ 10 kg/ha. Best results are obtained when the fields have adequate levels of available phophorus. In nitrogenous activity would account for daily input of approximately 0.5 kg nitrogen/ha.

TABLE 4

Effect of BGA Cultures on Available Phosphorus Content in N-free Bristol's Medium Incorporated with Non-water Soluble Inorganic Phosphorus

BGA cultures	Available phosphorus (mg/ml)	
	AP	TCP
Plectonema	8.00	3.20
Aulosira	12.56	11.60
Calothrix	0.56	0.90
Oscillatoria	8.46	6.80
Nodularia	0.40	0.36
Nostoc	0.45	1.30
S.E. +	0.11	0.09
CD (0.055 %)	0.24	0.20

AP = Aluminium phosphate,TCP = Tri calcium phosphate

Azolla–Anabaena symbiosis

The flooded water and the surface of the soil provide the sites for aerobic phototrophic nitrogen fixation by free living cyanobacteria and *Azolla Anabaena* symbiotic N_2 fixing complex. The *Azolla Anabaena* symbiosis fixes on an average of 60 kg N/ha. A great volume of work has been accomplished during the last 2-3 decades on *Anabaena azolla* symbiosis which include symbiotic biology of *Azolla Anabaena* mutualistic N_2 fixing complex. Preliminary meristem mutagenesis in *Azolla* for achieving reduce phosphate dependence, temperature tolerance and significant sporulation / spore germination under control conditions. The mass production biofertilizer technology of free living and symbiotic azolla anabaena interacting and agronomic effects of both kinds of cyanobacterial biofertilizers with rice and economics of rice cultivation with the cyanobacterial biofertilizers. While the research, so far, are an indicative of the strong potential of cyanobacterial biofertilizer technology in rice growing countries, open up vast area of more concerted basic, applied and extension work in future to make these self renewable natural nitrogen resources further promising at field level in order to health reducing the requirement of inorganic nitrogen to be the bare minimum, if not zero in biofertilizers technology for rice base cropping system.

Azolla is a tiny water floating fern and is having a blue green algal symbiont within a central cavity of leaf. The heterocyst of the symbiont *Anabaena* is the site of nitrogen fixation. The species of *Azolla* are *A. pinnata, A. nilotica, A. caroliniana, A. filiculoides, A. mexicana* and *A. microphylla*. *Azolla* occurs commonly in ponds, ditches and in rice fields of warm, temperate and tropical regions throughout the world. *Anabaena azollae* is the only species found in symbiotic association with *Azolla*. Due to application of *Azolla* 25 to 30 kg of N fixed/ha, besides an increase in paddy yields by 20 to 25 per cent and through its application there is an increase in organic matter content of soil leading to an increase in soil fertility.

Methods of application of *Azolla*

In India there are two recommended methods for applying *Azolla*. Application as green manure prior to a rice planting is the first method while second method involves dual cropping with rice, where *Azolla* grows simultaneously with the main crop.

Before transplanting rice seedling, the field prepared for *Azolla* inoculation by dividing it into subplots each of about 300 to 400 sq.m. in area. These plots are flooded with water, and inoculated with fresh *Azolla*. After 2 to 3 weeks, the water is drained out and *Azolla* is incorporated into the soil. In a week's time the rice seedlings are transplanted to the plots. Ideally, an optimum level of temperature and moisture has to be maintained, with adequate supply of nutrients especially phosphorus.

Azolla can not be advantageously used in areas that have no water, or, where pests and adverse temperature prevail. It is quickly perishable if removed from water and is thus not amenable to long distance transportation. *Azolla* finds use as a green manure, a fodder livestock, as a weeds suppressant and as composting material.

Acetobacter diazotrophicus : The India will need about 415 million tones of sugarcane by 2020 A.D. In view of speedily increase in the cost of chemical nitrogenous fertilizer, the marginal farmer due to his poor economical conditions can not afford the fertilizer to his crop with recommended high doses of chemical nitrogenous fertilizers which results into low productivity. Therefore, biological nitrogen fixation through micro-organisms has been found very economical and advantageous (Subba Rao, 1988).

Biological nitrogen fixation in sugarcane crop has been an

interesting area of research over several decades. N_2 fixing bacteria like *Azotobacter chroococcum* and *Beijerinkia* sp. were described as predominant nitrogen fixing organisms in root environment of sugarcane. After the discovery of *Azospirillum* spp. by Dobereiner and Day (1976) sugarcane has been investigated for nitrogen fixation by the associated nitrogen fixing bacteria. Recently *Acetobacter diazotrophicus* has been described as a potential diazotrophs in sugarcane (Cavalcante and Dobereiner, 1988). Scientists from countries like Canada, Mexico, Australia, Cuba and India have also confirmed the innate association of *Acetobacter diazotrophicus* with sugarcane.

Acetobacter diazotrophicus play important role in nitrogen economy of sugarcane and it also reduces the incidence of red rot disease in sugarcane. *Acetobacter diazotrophicus* is cheaper biofertilizer and it will help to increase yield level of sugarcane crop. It's production technology is simple as well as the method of application is easy. The use of biofertilizer *Acetobacter diazotrophicus* will definitely help to increase the productivity level in sugarcane.

Acetobacter diazotrophicus biofertilizer is a recently recognized endosymbiotic nitrogen fixing bacterium gaining much importance in sugarcane. It is widely accepted and sugarcane farmers appreciated the performance of this biofertilizer. A praise worthy and supervising feature of this bacterium is that it is abundantly present in the xylem sap of sugarcane stem, leaf, leaf-sheath and root tissue. Cultures of *Acetobacter diazotrophicus* can be isolated easily from the root tissue of sugarcane or from the juice of crushed sugarcane using nitrogen free semisolid LGIP medium. As compared to India, Maharashtra State contribute 30-50 % of production of sugarcane.

Collection of various genotype setts the stem portion washed thoroughly in running tap water to remove the dirt particle rind of cane sterilized by $HgCl_2$ solution (0.1 %) for few second subsequently 70 % ethanol used. Rinsed with sterilized water to remove traces of ethanol and $HgCl_2$ crushed sugarcane and juice collected in beaker. Prepare serial dilution upto 10^{-8}. Take 1 ml of suspension from 10^{-6} fold dilution place in a sterile petriplate and poured semisolid LGIP media on it. Incubate petriplates at 28° ± 2°C, temp. for 10 days. Orange colonies are formed on LGIP medium. These colonies are purified on modified liquid potato medium. Then individual colonies are transferred to LGIP slants. The individual colonies are small, dark brown, raised, circular dense. Further mass multiplication of *Acetobacter diazotrophicus* done in LGIP broth culture.

Acetobacter diazotrophicus :

- It is a gram negative, non spore forming, microaerophilic, diazotrophic, endosymbiotic bacterium.
- It is short straight rod with round ends and measures 0.7 to 0.9 m in length and 0.2 m in width.
- Cells are motile with either a single or three lateral flagella.
- It has an optimum growth temperature of 30 °C.
- The optimum pH for the bacterium is 4.0 to 4.5.
- On LGIP medium the colonies of *Acetobacter diazotrophicus* appear irregular 2-3 mm diameter smooth, flat with bright yellow or yellow orange colour.
- On potato dextrose agar medium (10 % glucose) the colonies could produce dark brown pigment.
- *Acetobacter diazotrophicus* is more sensitive to antibiotics like rifampicin and tetracyclines and fairly resistant to streptomycin.
- It survives in soil poorly.

Role of *Beijerinckia, Derxia* and other minor diazotrophs in improving soil fertility

Soil fertility is a condition of soil having enrichment of most of the micro as well as macro element which have capacity to give nutrient rich media for the optimum growth of plants.

Soil fertility includes enrichment of all essential 16 elements for plant growth. Nitrogen is one of the key element required to proper growth and maturity of plants. Nitrogen plays an important role in optimum growth of plant to keep the essential osmotic pressure and to develop the tissue of plants and organs of cell.

As soil contain nitrogen in more amount than air, it is essential to add the nitrogen through various means especially for getting the higher yield. Though there is some natural source to get nitrogen like lightning, it deposit very less amount. As the trend in agriculture are getting changed, it become necessary to apply nitrogen by various way other than the natural.

Now a days the chemical and biological fertilizers are on the way of success. It is so because it gives required amount of nutrient at proper time and quantity. It is also available widely and easy to apply. As considered the chemical fertilizer, it comes to know that, nutrient response is rather lower than that of hazardous nature of it for both i.e. plants as well as animals. Hence, the revolution of the biofertilizers is coming on the way.

As considered the biological fertilizers, though they have long time to take for action, but it is eco-friendly having no residual effect or other side-effects. As we see former that the nitrogen is a prime source for plant growth, it is become essential to know the micro-organisms which play the important role in synthesis of nitrogen.

Though there is a vast known about most of nitrogen fixing organisms such as *Rhizobium, Azotobacter, Azospirillum,* Blue green algae, there is very little record about minor diazotrophs like *Beijerinckia, Derxina, Frankia, Bacillus, Arthobacter,* etc. Hence, it is essential to give focus on that point to get use of this minor diazotrophs for innovating new biofertilizer.

Nitrogen is fixed by diazotrophs in nitrification process i.e. biological oxidation of ammonia to nitrite by *Nitrosomonas* bacteria and nitrite to nitrate by *Nitrobacter* in aerobic condition. This nitrate become available as a source of nitrogen for absorption through roots of plant.

As we study the importance of minor diazotrophs and the way of nitrogen fixation by them, we have to know a little about this minor diazotrophs. The diazotrophs we are going to study are *Beijerinckia, Derxia, Clostridium, Klebsella, Frankia, Arthobacter and Streptomyces.*

1. Beijerinckia

Occurrence : *Beijerinckia* is occurred mostly in tropical zone laterite soils. This is because it is slow growing organism and effectively compete with other soil microbes under optical physical and chemical condition i.e. 30 - 35 °C temperature and acidic conditions i.e. pH ranges from 5.0 to 6.5.

Requirement : *Beijerinckia* does not require calcium for their effective growth. They require iron in higher amount and magnesium as well as phosphorus in lower amount. They are strictly inhibited by calcium carbonate. They have ability to tolerate element as aluminium and manganese. They require moist condition for good growth but at unavailability they can produce 'zoogloeae' and survive for long period in air dried condition.

Role in soil : :

1. All the strains are capable of fixing molecular nitrogen which has a sole source obtained from oxidation of ammonium and nitrate but not the glutamate.
2. It fixes 1 gm of nitrogen / gram of carbohydrates. High aeration rate inhibit the rate of nitrogen fixation.
3. Under aerobic condition they produces acid from carbohydrates and utilize for growth.
4. They oxidize gluconate and lactinate in soil.
5. They have capacity to produce polysaccharides from sucrose and raffinose.
6. They produces non-diffusable orange pigment in presence of iron but in absence of benzoate.
7. They are well producers of catalase and oxidase.
8. Some spp. of *Beijerinckia* also responsible to reduce nitrate to nitrite.
9. Some spp. of *Beijerinckia* are responsible for hydrolysis of urea in soil.

2. Derxia

Derxia is a minor diazotrophs and have required slightly acidic pH i.e. 4.6 – 6.5 for their optimum growth. Their occurrence is more in flooded soils which are present at tropic. In India they are mostly occurred in field of Jute and Forage grasses.

Role in Soil :

1. *Derxia* are mesophillic in nature. They are aerobic micro-organisms and produce acid from carbohydrate.
2. They have ability to fix lower amount of nitrogen and fix 25 mg of nitrogen per gram of carbon source
3. They produce yellow brown pigment in soil required for better structure of soil particles.
4. They require presence of *Molybdenum* for fixation of atmospheric nitrogen.
5. *Derxia* have wide range of utilization of sugar, alcohol, organic acids and to oxidize it to carbon di oxide.

6. They have also ability like *Beijerinckia* to hydrolyse the urea in soil.
7. They have unique ability to produce the indigocine in soil.
8. They produce some enzymes like catalase and oxidase which play important role in formation of soil structure.

3. Frankia

Frankia is a nodule symbiont of large number of non-legume crops. The nitrogen activity is associated primarily with vesicles. The vesicles can be formed terminally or laterally borne.

4. Streptomyces

Streptomyces is aerobic chemo-organotrophs having oxidative type of metabolism. They catalise positively nitrate to nitrite and degrade adenin, esulin, casein, gelatin etc. Wide range of organic compounds are used as sole source of carbon for energy and growth.

5. Bacillus

Bacillus is a large gram +ve rod shaped bacteria, which bears spores with a tendency to form chains and capsules. Some bacilli have flagella and are motile. Bacilli are ubiquitous in soil and air, and are often responsible for food spoilage. This group includes few N fixing bacteria. In *Bacillus* spp. oxygen is terminal electron acceptor in respiratory metabolism. In some species nitrate can be utilized as alternate electron acceptor and denitrification occurs under oxygen limiting condition.

6. Clostridium

Clostridium have ability to formulate many organic compounds. They contain many enzymes responsible for break down of large molecules and have important role in bio-degradation and carbon cycle. Some of the species of *Clostridium* fix nitrogen independently from plants.

7. Arthobacter

Arthobacters decrease with increasing soil acidity. Mostly they are found in sediment sewage. *A. celobiformis* forms symbiotic relationship with *Anabena* and *Azolla*. They use choline oxidase to prevent denaturation of soluble enzyme at high salt concentration.

ROLE OF COMPOST CULTURES IN ORGANIC FARMING

Micro-organisms play a vital role in the environment. Without the action of micro-organisms, dead animal and plant residues would have accumulated to such an extent that they would have choked the surface of earth. The conventional methods of producing compost from plant residues takes long time and the compost prepared by these methods is low in nutrient content. The best alternative lies in evolving rapid method of composting consistent with improving the nutrient content of the final product. As the plant ages, the content of water-soluble constituents, proteins and minerals decreases and the percentage in abundance of cellulose, hemi-cellulose and lignin rises. These substrates constitute the mixed and highly diverse substrates utilized by the soil microbial population in mineralizing carbon compounds.

1. Cellulolytic micro-organisms

A prominent carbonaceous constituent of higher plants and probably the most abundant organic compounds in nature is cellulose. Because a large part of the vegetation added to soil is cellulosic, the decomposition of this carbohydrate has a special significance in recycling of nutrients. The cellulose utilizing population includes aerobic and anaerobic mesophilic bacteria, filamentous fungi-basidiomycetes, thermophilic bacteria, actinomycetes and certain protozoa. A diverse group of fungi utilizes cellulose for its carbon and energy sources. The list of such cellulolytic microbes is given in Table 5.

TABLE 5

Cellulolytic Microbial Genera

Fungi	Bacteria	Actinomycetes
Alternaria, Aspergillus,	*Achromobacter,*	*Micromonospora, Nocardia,*
Chaetomium, Coprinus,	*Angiococcus,*	*Streptomyces,*
Fomes, Fusarium,	*Bacillus, Cellulomonas,*	*Streptosporangium.*
Penicillium, Polypores,	*Cytophaga,*	
Rhizoctonia, Rhizopus,	*Sporocytophaga,*	
Trametes, Trichothecium,	*Pseudomonas,*	
Verticillium,	*Vibrio, Clostridium.*	
Zygorhynchum.		

2. Acceleration of composting by micro-organisms

Composting involves microbiological processes, which go on naturally with the help of native cellulolytic micro-organisms in compost pits. A question has often been asked whether addition of laboratory grown cultures of cellulolytic micro-organisms enhance the rate of microbiological processes in composting. By addition of selected cellulose decomposers accelerate the microbiological activities in composting resulting in the alteration in C : N ratio, loss in weight and humic acid contents to the beneficial side. The rate of decomposting culture is ½ kg/ton of organic waste. In 3 to 4 months the compost is ready, normally it takes 8 to 10 months for compost preparation.

POTENTIAL ORGANIC WASTES FOR COMPOSTING

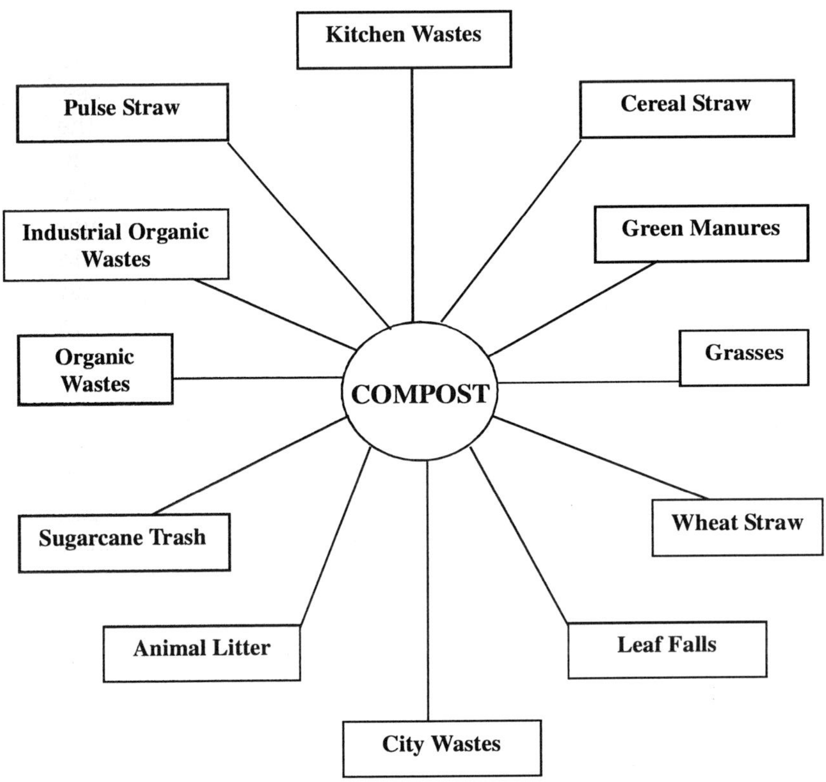

ROLE OF PHOSPHATE SOLUBILIZING BIOFERTILIZERS IN ORGANIC FARMING

Biofertilizers are the fertilizers prepared from bio-agents containing especially microbial cells. These are the microbial preparations containing live cells of bacteria which fix atmospheric nitrogen, solubilize insoluble phosphate in soil or decompose cellulose organic wastes.

Importance of phosphorus in plant growth

Phosphorus is an important nutrient to plant next to nitrogen. Phosphorus is considered as key to the crop productivity. It is associated with many vital functions and responsible for several plant growth characters such as utilization of sugar and starch, photosynthesis, nucleus formation and cell division, fat and albumin and also for formation, cell organization and heredity transfer. It stimulates growth of roots. It helps in early maturity of crop.

Importance of phosphate solubilizing biofertilizers

All India production of straight phosphate is 32,81,500 metric tones. Consumption of phosphate in India in the year 1999-2000 was 48,04,050 metric tones and in Maharashtra consumption of phosphate is 5,51,950 metric tones (Anonymous, 2002).

Most soils contain insoluble and inorganic phosphates, but these are of no use to the crops unless they are solubilized. Also, though soils contain organic phosphorus, it can be used by crops only if it is mineralized.

Various soil micro-organisms play important role in solubilizing insoluble inorganic and mineralizing organic phosphorus compound and these microbes are commonly called phosphate solubilizing micro-organisms. Several soil bacteria particularly belonging to *Pseudomonas, Bacillus* and *Aspergillus* possess the ability to bring insoluble phosphate in soil in to soluble forms by secreting organic acids such as formic acid, acetic acid, fumaric acid and succinic acid. These acids lower the pH and bring about the dissolution of bound forms of phosphate (Subba Rao, 1993). The phosphate solubilizers also produce phyto-harmones and growth promoting substances.

Microbes responsible for phosphate solubilization

Phosphate solubilizing micro-organisms include various species of bacteria, fungi, actinomycetes and yeasts. The list of these micro-organisms is given in Table 6.

TABLE 6

Phosphate Solubilizing Organisms

Type	Examples
Bacteria	*Bacillus megatherium* var. *phosphaticum, B. megatherium* var. *serratia, B. polymyxa, B. circulaus, Pseudomonas liquifaciens, P. striata, Achromobacter* spp., *Arthrobacter* spp.
Fungi	*Aspergillus niger, A. fumigatus, A. flavus, A. awamori, Penicillium digitatum, P. liliacinus, Cephalosporium* spp. *Trichoderma* spp.
Actinomycetes	*Streptomyces* spp., *Nocardia* spp.
Yeasts	*Rhodotorula* spp., *Schwanniomyces occidentalis*

Of the above listed phosphate solubilizing micro-organisms bacteria *viz., Bacillus polymyxa* and *Pseudomonas striata* and fungi *viz., Penicillium digitatum* and *Aspergillus awamori* are used in commercial preparations.

Mechanism of phosphate solubilization

The mechanism of phosphate solubilization by soil micro-organisms includes :

(i) Formation of organic acids

(ii) Formation of phosphatase enzymes

(iii) CO_2 evolved due to respiration and other activities of microbes from carbonic acid in soil solution which solubilizers insoluble phosphates in soil and

(iv) Indirect effect of sulphur oxidizers and nitrifiers in soil leading to formation of sulphuric and nitric acids, respectively which solubilize inorganic bound phosphatic compounds.

The major mechanisms is production of organic acids. The organic acids produced by phosphate solubilizing micro-organism are listed in Table 7.

Table 7

Acids Produced by Phosphate Solubilizing Micro-organisms

S.No.	Type	Examples
1.	Monocarboxylic acids	Acetic and Formic acids
2.	Monocarboxylic hydroxy acids	Lactic, Gluconic and Glyconic acids
3.	Dicarboxylic acids	Oxalic and Succinic acids
4.	Dicarboxylic hydroxy acids	Malic, Maleic and Tartaric acids
5.	Tricarboxylic hydroxy acids	Citric acid

Phosphate solubilizing bio-inoculants/biofertilizers

Phosphate solubilizing bio-inoculants/biofertilizers are prepared from the bacteria or fungi which solubilize fixed form of phosphate in soil and make it available to plants. Phosphatic bio-inoculants are known by various names *viz.*, phosphobacterin, microphos, bacteriophos, phosphomicrobial biophos, etc. Phosphate solubilizing bio-inoculants are prepared either as carrier based inoculants from phosphate solubilizing bacteria or as sorghum based inoculant from phosphate solubilizing fungi.

Commercially produced phosphate solubilizing bio-inoculants

In USSR, a commercial biofertilizer under the name "Phosphobacterin" was first prepared by incorporating *B. megaterium* var. *phosphaticum* and widely used in USSR and other east European countries with yield increase of the order of 5-10 per cent over corresponding controls. Carrier based IARI microphos culture has been developed from efficient strains of bacteria *viz., Pseudomonas striata, Bacillus polymyxa* and fungi *viz., Aspergillus awamori* and *Penicillium digitatum*. Biological Nitrogen Fixation Scheme of Mahatma Phule Krishi Vidyapeeth located at College of Agriculture, Pune is manufacturing two types of phosphate solubilizing biofertilizers *viz.*, i) carrier (lignite based) and ii) Sorghum grain based phosphatic biofertilizers.

Importance of phosphate solubilizing bio-inoculants

The efficient phosphate solubilizing bio-inoculants have shown capacity to solubilize insoluble inorganic phosphate such as rock phosphate, tricalcium phosphate, iron phosphate and aluminium phosphate in soil. These phosphatic bio-inoculants also mineralize organic phosphatic compounds present in organic manures and in soil resulting in availability of P to field crops. The results of large number of field trials invariably showed that inoculation of crop seeds with P solubilizing bio-inoculants increased the grain and dry matter yields, protein contents of grains and P uptake by crops inoculated indicating the solubilization of fixed soil phosphorus. In legume crops along with *Rhizobium* bio-inoculants, the phosphate solubilizers are more promising. They help in increasing activities of *Rhizobium* as P is necessary for diazotrophy of *Rhizobium.*

Potash mobilizing micro-organisms

The readily exchangeable fraction of the potassium is believed to be present especially on the surface of the clay components, so that the finer this clay the greater the amount of this fraction. The non-exchangeable fraction is believed to reside in those expandable constituents of the clay which 'gorge' the potassium into the interstices between the crystal layers and thereby trap the potassium. Use of K-bacteria (*Frateuria aurentia*) into the clay is believed to release some of this gorged potassium. The Potash Mobilizing Bacteria is a gram negative rod type bacteria, can grow about pH 3.5-11 and capability to mobilizing the mineral potash in the tone of 40-60 kg/ha. As the potash is producing the organic acids, therefore, combination of the potash bacteria in the formulation was quite typical.

The main function of K is the activation of numerous enzyme systems involved in the formation of organic substances and in the build -up of compounds such as starch or protein. K is involved in cell enlargement and in triggering the growth of young tissues. K improves the water status of the plant and water use efficiency in general(Sahoo,2002). K promotes photosynthesis and the transport of the assimilates to the storage organs. K improves the tolerance of plants to stress. The physiological functions of potassium are of great practical significance in plant growth and horticulture. K is essential for the development of the root system (Ramarethinam et al., 2006) K increases the sugar content of crops such as fruits, carrots and sweet potatoes.K increases the size of fruits. K improves the colour of fruits and flowers. K improves the keeping quality of fruit and vegetables. K is essential for the efficient nitrogen fixation by leguminous crops.

Actinomycetes

Actinomycetes are filamentous organisms that are intermediate between bacteria and fungi. They have a mass of fine elongated threads, called mycelium, which resembles the typical growth structure of fungi. They reproduce asexually by fission, or through spores.

Actinomycetes come under the bacterial order of actinomycetales. Like fungi, actinomycetes secretes some polysaccahrides and help water to be retained in soils. They are distinct from fungi in many respects. Their cell wall is typical of bacteria. They are prokaryotic, which means that their genetic material is not enclosed in the cell nucleus.

Most actinomycetes are aerobic and help decompose organic matter like cellulose and other resistant organic molecules. Very few of the actinomycetes establish symbiotic nitrogen fixing association with plants like soapberry, alders, bitterbrush, coffee berry etc.

Actinomycetes are helpful to atleast seven botanical families, whereas rhizobia bacteria help only one family *viz.*, leguminosae. Actinomycetes synthesize a great variety of antibiotics like streptomycin, aureomycin, terramycin and neomycin.

Acinomycetes develop best in moist and well aerated soils through they can also survive in soils with low moisture level. Being aerobic, they participate in the breakdown of organic matter to release plant nutrients. The characteristic odour freshly ploughed moist field is due to soil actinomycetes.

ROLE OF BIOLOGICAL AGENTS IN ORGANIC FARMING

Trichoderma spp.

It is an established fact that *Trichoderma* spp. are capable of effectively controlling many soil-borne plant pathogens because of being the main competent in the rhizosphere of crop plants. *Trichoderma,* similar to all fungi lack chlorophyll, possesses heterotrophic mode of nutrition requiring preformed organic matter and hence it is the main competitor for organic nutrients in the rhizosphere of crop plants. *Trichoderma* spp. have ability to adopt extreme soil conditions *viz.*, soil temperature, moisture and fungicidal load added in the soil and that is why, it is gaining importance in the field of biological control of plant diseases.

Trichoderma has conidiophores erect, highly ramified, more or less conical, verticullate bearing flask shaped phialides, singly or in clusters. Conidiophores bear subglobose or ellipsoid, slimy, nonseptate phialospores. These phialospores are often gathering in balls at opening of phialides. Chlamydospores are intercalary, globose, ellipsoidal, hyaline, smooth walled, colonies grow rapidly. *Trichoderma* has become a model biocontrol agent in the management of plant disease because of the following characters.

1. *Trichoderma* is easy to isolate and culture.
2. It is ubiquitous.
3. It grows rapidly on many substrates.
4. It affects a wide range of plant pathogens.
5. It is nonpathogenic to higher plants.
6. *Trichoderma* suppresses pathogens by all three mechanisms i.e. antibiosis, mycoparasitism and competition.

Methods of application

The methods presently used for effective management of diseases includes the application of *Trichoderma* by any of these methods *viz.*, Broadcast application, furrow application, root zone application, seed coating or seed treatment and wound application.

(1) Broadcasting : Broadcasting is used for soil application. The quantity required is high and varies from 125 – 250 kg/ha. It is mixed with FYM and broadcasted on the soil and then incorporated into soil through harrowing. It is used to control *Fusarium* wilt, *Sclerotium* foot rots and *Macrophomina* root rots. *Trichoderma* is applied by this method at least 2 weeks before sowing of crops.

(2) Furrow Application : It is comparatively economical to broadcasting and is followed in nurseries. The quantity required 130-160 kg/ha. It should be applied in the open furrows covered with the soil. *Trichoderma* is applied by this method 15 days before sowing of the seeds.

(3) Root Zone Application : This is used in wide spaced crops such as plantation and fruit crops. It is done in peach and plum against silver leaf disease.

(4) Seed Treatment : In this method of application, the seeds are first mixed with sprinkles of water 5-7 ml/kg so that the seed surface becomes moist. Take known quantity of seed formulation (Powder form) 4-5 g/kg seed, mix thoroughly by agitating the container so that the powder adheres to the seed surface. Then treated seeds should be dried in shade for two hours and sown immediately after treatment. This formulation can be mixed with sand @ 2.5 kg formulation in 50 kg sand and this mixture can be used to broadcast around the roots and mixed with the soil. This helps good seed germination, seedling emergence and finally good plant stand. It also gives protection in the rhizosphere against root rots and wilt of various crops such as groundnut, bengal gram, sunflower, tobacco, sorghum, cotton, red gram, green gram, black gram and rice caused by *Fusarium*, *Macrophomina* and *Sclerotium*.

Gliocladium spp.

Gliocladium is widely distributed all over the world and occurs in nearly all soils and other natural habitat specially those consisting of organic matter. The abundance of *Gliocladium* in various soils, coupled with their ability to degrade various organic substrates in soil; their metabolic versatility and their resistance to microbial inhibitors, suggest that they may possess the ability to survive in many ecological reaches depending on prevailing conditions and the species or strain involved. The ability of *Gliocladium* to act as mycoparasites of hyphae and resting structures of plant pathogens has been demonstrated *in vitro* and in natural soil. *Gliocladium virens* is reported to produce antibiotics *viz;* Gliotoxin, Gliovirin and viridin.

Pseudomonas fluorescence

Plant growth promoting rhizobacteria (PGPR) in combination with efficient rhizobia, can affect the growth and nitrogen fixation in various pulse crops by inducing the occupancy of introduced *Rhizobium* in the nodules of legume. The production of phytohormones by PGPR is consider to be an important mechanism by which these bacteria promote plant growth.

Pseudomonas fluorescence are effective bacterial antagonist for the management of soil borne and foliar diseases. Seed treatment plus soil application of talc-based formulation of *P. fluorescence* effectively reduces the disease incidence. Expression of various defense related enzymes and chemicals were found involved in the induction of systematic resistance against pathogen infection.

Characters

Pseudomonas is straight or slightly curved rod 0.5 - 1.0 mm in diameter and 1.5 - 5.0 m in length. Many species accumulate poly-b-hydroxybutyrate as carbon reserve material, which appears as sudanophilic inclusion. Do not produce prosthecae and are not surrounded by sheath. No resting stages are known. They are Gram-negative and motile by one or several polar flagella. Aerobic having a strictly respiratory type of metabolism with oxygen as the terminal electron acceptor, in some cases nitrate can be used as an alternate electron acceptor. Most species, if not all, fail to grow under acidic condition (pH 4.5). Most species do not require organic growth factor and are oxidase positive chemolithotrophs; able to use H_2 or CO_2 as energy source; widely distributed in nature. The mol. % G + C of the DNA is 58 - 78 (Bd). Early treatment of the genus *Pseudomonas* included pigmentation as a generic character, but at present, the genus includes many so called non-pigmented species e.g. *P. stutzeri*. Actually, cellular component, which in some cases become quite apparent. The colonies of many strains become reddish brown with age due to high concentration of cytochrome C in the cell.

Among most of *Pseudomonas* the soluble pigments are the fluorescent (Pyovedin) and pyocyanin. Fluorescent pigments are produced abundantly in media of low iron concentration and the fluorescent varies from white to blue green upon excitation with UV radiation. Kloepper *et al.* (1980) described a fluorescent chelator under the name Pseudobactin to which they attribute the role of promoting plant growth by inhibition of plant pathogenic bacteria in the rhizosphere of some cultivated plants.

Pseudomonas fluorescence, earlier known as *Bacillus fluorescence* requires optimum temperature of 25 - 30°C for growth. Found in soil and water from which it can be isolated after enrichment in media containing various carbon sources. Incubated aerobically, strains of the denitrifying biovars can be enriched in similar media containing nitrate and incubated under anaerobic condition.

Habitat of *P. fluorescence*

Fluorescent *Pseudomonas* sp. can be isolated from rhizosphere of different crops from phylloplane, from farm wastes, diseased parts of plant and stagnated field water.

In the microbioal society the antagonist bacterium like *Bacillus thermophillic* (Borkar *et al.*, 1999), *P. fluorescence* and fungi *Trichoderma* play important role to maintain the social colonization of microbes. Among the naturally occurring population of fluorescent *Pseudomonas* and *Trichoderma harzianum* in rhizosphere had high antagonistic potential against several plant pathogens. However, *P. fluorescence* is highly inhibitory to *Trichoderma harzianum* (Varsheneya *et al.*, 2000).

Role of *P. fluorescence* as biopesticide

Fluorescent *Pseudomonas* suppress the pathogens by various mode of actions namely competition for nutrients and space, antibiosis by production of various antibiotics, siderophores and lytic enzymes. Other mechanisms include production of hydrogen cyanide and degradation of toxins (Defago *et al.*, 1990). Moreover, fluorescent *Pseudomonas* produce plant growth-promoting substances such as auxins and gibberellin and enhance plant growth and yield. Hence they are collectively called as Plant Growth Promoting Rhizobacteria (PGPR) (Dubeikovsky *et al.*, 1993). In addition to direct antagonism against pathogen and plant growth promotion, induce systemic resistance (ISR) by enhancement of plant defence capacity. Fluorescent *Pseudomonas* is a new strategy for plant disease management (Ramamoorthy, *et al.*, 2001). Fluorescence *Pseudomonas* activates ISR against various fungal, bacterial and viral diseases (Lin *et al.*, 1995 a, b) and nematode pest. Application of fluorescent *Pseudomonas* as seed treatment alone or as seed treatment + soil application has induced systemic resistance (Wei *et al.*, 1996).

Colonization of beet root by fluorescent bacteria has correlation with induction of pathogenesis related proteins (PR proteins) and increased activity of PAL. Maurhofer *et al.*, (1994) reported that ISR by *P. fluorescence* strain CHAO against tobacco mosaic virus in tobacco was associated with accumulation of b-1-3 glucanase and endochitinase. In tomato *P. fluorescence* strain 63-28 induced the accumulation of chitinase and prevented the infection of *Fusarium oxysporum* f. sp. *radicis lycopersici* (M'piga *et al.* 1997).

These bacteria produce indole acetic acid (for plant vigour), increase the amount of phenolic (act as a fungal toxicant), increase the PAL activity (PAL is the first enzyme in phenyl propanoid metabolism) and play as important role in the biosynthesis of phenolics and phytoalexins (Daffy *et al.*, 1997). The activities of b-1-3 glucanase and chitinase is higher in bacterized plant challenged with pathogen and higher activities are observed at 3 – 5 days after challenge inoculation. ISR by fluorescent *Pseudomonas* involves strengthening the physical and mechanical strength of the host cell wall and causing biochemical and physiological changes leading to synthesis of defense proteins (Vishwanathan and Samiyappan,1990;Maurhofer *et al.,* 1998).

Accumulation of phenolics, PAL, PO, PPO, b-1-3 glucanase and chitinase by *P. flurescence* isolate pf1 in chilli leaves have collectively contributed to induce resistance in chilli leaves against *Colletotrichum capsici* (Ramamoorthy and Samiyappan, 2001).

P. fluorescence and *P. putida* suppress several major plant pathogens. The former is well known for its usefulness in the biocontrol of "take all" disease of wheat caused by *Gaeumannomyces graminis* var. *tritici* (Cook and Rovira, 1976). The suppression depends on the ability of the bacteria to colonise the roots and production of an antibiotic phenazine-1-carboxylic acid (PCA), a siderophore called pyoverdin and an antifungal factor (AFF). Iron regulated non-siderophere antibiotics may be produced by fluoroscent *Pseudomonas* more frequently than previously recognized and could be partially responsible for beneficial effects that were attributed in the past to fluorescent siderophores (Thomashow and Weller, 1996). A strain of *P. fluorescence* CHAO isolated from swiss soil naturally suppressive to tobacco black root rot caused by *Thielaviopsis brassicola* suppressed the disease in iron sufficient soils (Stutz *et al.,*1986).

The bacterium produced pyoverdin and several secondary metabolites with antibiotic properties e.g. HCN, pyrrolnitrin and 2 – 4 diacetyl phloroglucinol (DAPG). Siderophores produced by *Pseudomonas* species have been implicated in the biological control of damping off of cotton (Loper, 1988; Laha *et al.*, 1992), root rot of wheat (Backer and Cook, 1988) caused by *Pythium ultimum*, potato seed piece decay by *Erwinia carotovora* (Xu and Cross, 1986) and in suppression of several vascular wilts caused by *Fusarium oxysporum* (Kloepper *et al.,* 1980 and Sneh *et al.,* 1984).

Bacillus spp.

Bacillus is spore forming, rod shaped, crystalliferous, Gram +ve bacterium belonging to Prokaryotic Kingdom, Firmicutes Division, Firmibacteria Class and Bacillaceae family. The important species are *Bacillus subtilis, B. polymixa* and *B. thuringenesis. Bacillus subtilis* is reported to produce antibiotic Bulbiformin.

ROLE OF SIDEROPHORE IN ORGANIC FARMING

Most aerobic and facultative unaerobic micro-organisms respond to low-iron trees by producing extracellular, low molecular-weight (500-1000 daltons) iron (III) transport agents, designated siderophores, with selectively complex iron (III) with very high affinity (John Leong, 1986).

Although iron is abundant in soil (1-6 %) and it ranks fourth among all elements on surface of earth, however, it is often unavailable to micro-organisms and plants because of it's solubility which is dependant on pH, under aerobic soil conditions, most of the iron exists in insoluble ferric form and tends to precipitate to form polymer of oxyhydroxide-FeOOH. Iron plays an important role in micro-organisms and plants. This is a component of cell and it's deficiency can cause growth inhibition, decrease in RNA and DNA synthesis and inhibits sporulation and can also change the cell morphology. Iron exists in two oxidation states (Fe^{2+} and Fe^{3+}) and therefore, it is suitable as an electron transporter. In intermediatory metabolic process, iron is required in TCA cycle, electron transort chain, oxidative phosphorylation, nitrogen fixation, aromatic biosynthesis and photosynthesis. Iron is also known to regulate the biosynthesis of porphyrin, toxins, vitamins, antibiotics, cytochromes, pigments, siderophores and aromatic compounds. Iron is present as co-factor or required by different enzymes and proteins such as peroxidase, superoxidase dismutase, nitrogenase, hydrogenase, glutamate synthatase, ribonucleotase, diphosphate reductase, aconitage, DAHP synthatase, cytochromes, ferridoxin and flavoproteins. Iron storage proteins like ferritin in animals and bacterioferritin in micro-organisms have also been discovered.

To meet the requirement of iron, micro-organisms and plants must acquire an adequate supply of iron. Therefore, living organisms have evolved a specific and high affinity mechanisms to acquire iron (Dudeja *et al.*, 1977).

Some micro-organisms produce iron-scavenging molecules. The predominant competition for Fe^{3+}, in the rhizosphere seems to be between different species or strains of bacteria. This may either take the form of direct competition between a species which produces an iron-scavenging siderophore and species which do not have such a capacity or probably more likely, between two different types of siderophore molecules produced by two competition species. The competition will then depend on the relative affinities of the two types of siderophore molecules for the Fe^{3+} ions, but also on the capacity of the bacterial cell surface and membrane complex to release the bound iron from the siderophore molecule. Thus, the successful species in such a competition will be either the species which produces the siderophore molecule with the highest affinity for Fe^{3+} ions or the species which is able to release the iron not only from it's own siderophore molecules but from those produced by one or more of it's competitors. In this way some species of bacteria may be able to pivate Fe^{3+} ions cheleted by siderophores in the soil which originated from competior species (Schippers *et al.*, 1987).

Uptake of Iron

It is likely that competition for limited minerals play a significant role in the rhizosphere. Competition for iron has been the most studied case. Most of the root will be in an oxidizing environment where iron will exist in the extremely insoluble Fe^{3+} form. To ensure it's own growth, each organism must convert part of the iron in root region into a form to which it has access and (if that form is inaccessible to competing organisms) the iron acquisition mechanism can also function to deny this essential material to a competitor. In this way, a single mechanism may serve two quite distinct functions. It is, however, possible that a competitor may serve two quite distinct functions. It is, however, possible that a competitor may evolve a receptor system capable of binding the ferrated form of the siderophore and thus benefit from this system. Such "Parasitism" of iron uptake mechanism is shown by *E. coli* and relatives; which can take up iron from ferrated ferrichromes. Ferrichromes are fungal siderophores (Hedges and Messens, 1990).

Mechanism

Typically an organism produces a repressor which regulates it's iron uptake mechanism, when there is no shortage of metal the repressor will be active (presumably, bound iron acting as corepressor). If the cell is starved of iron, the derepressed genes are expressed, producing enzymes required for the synthesis of a siderophore (a small soluble molecute capable of chellatin, Fe^{3+}). The siderophore is actively secreted and diffuses into the environment. It can bind ferric iron so powerfully as to remove this from most inorganic complexes likely to be found in soil. The ferric siderophore complex is recognized very specifically by a receptor in the cell membrane (itself the product of gene and genes regulated by iron) of the siderophore producer.

Ecology and uptake modulation

The siderophores of different organisms are a remarkably diverse set of molecules, presumably implying that iron uptake mechanism have evolved independently on many different occasions. It is clear that certain types of siderophores are particularly well adopted to particular ecological situations. *Azotobactor vinelandii* is highly adopted to exploiting the iron resources of soils. Untressed it forms a cell wall able to complex iron (which can be used by the cell) and produces a weakly binding chelator (with reductive properties) when relatively short of iron it produces azotochelin (a more complex chelator) and under extreme iron shortage a third (and apparently most efficient) chelator, azotobactin plus membrane proteins probably required.

TABLE 8

Partial List of Siderophores in Microbial Species (Neilands, 1984)

Organism	Siderophore (s)
Bacteria :	
Enteric spp.	Enterobactin, aerobactin
Paracoccus denitrificans	Parabactin
Vibrio cholerae	Vibriobactin
Bacillus megaterium	Schizokinen
Anabaena spp.	Schizokinen
Azotobacter vinelandii	E-bis-2,3-dihdroxybenzoyllysine
Pseudomonas spp.	Pyochelin, Pyoverdine, Pseudobactin, Ferribactin
Actinomyces spp.	Ferrioxamines
Mycobacteria	Mycobactins
Fungi :	
Widely distributed in Ascomycetes and Basidiomycetes	Feerichromes, Coprogen
Rhodotorula spp.	Rhodotorulic acid
Ectomycorhizol sp.	Hydroxymate type

BIOFERTILIZERS AS BIOCONTROL AGENTS

Soil is a heterogeneous habitat of beneficial, plant pathogenic and saprophytic microbes. The beneficial activities of soil microbes include production of growth promoting substances which tolerate growth of plant roots, degradation of toxic chemicals in soil, recycling of nutrients mainly through decomposition of organic residues in soil and making the availability of nutrients to crop plants. The commercial preparations of performing these useful activities are known as biofertilizers. In contrast, harmful activities of soil microbes especially soil-borne plant pathogens include damaging root tissues, exerting toxic effect on crop roots through production of toxins or cause severe melodies like rotting, wilting, etc. resulting severe crop losses.

Biocontrol

The control of soil-borne plant diseases is very difficult by conventional chemical applications. Secondly, many of these chemicals are not biodegradable easily and hence become toxic to beneficial living micro-organisms in soil. With the increasing awareness of conserving natural resources and reducing air, water and soil pollution, emphasis has now been given on biocontrol of soil-borne plant pathogens. During the last 30 years, there has been uncountable reports indicating and appreciating the manifold benefits from biofertilizers and one of these is protecting plant against soil-borne plant pathogenic fungi, viruses and nematodes. Use of biocontrol agents, which minimize use of hazardous chemicals is a major thrust area in present day strategy for management of plant diseases. Any condition or practice under which activity of pathogen is retarded through the agency of living organisms (including man himself) which results in reduction in the incidence of diseases caused by the pathogen. In simple words, biocontrol of plant diseases means management of plant diseases by biocontrol means of biotic agents.

MECHANISM OF CONTROLLING PLANT DISEASES

Production of antifungal compounds

The beneficial effects of *Azotobacter* are not only due its ability to fix nitrogen but also due to its capacity to produce anti-fungal antibiotics, which improve plant stands in inoculated fields (Lakshmi-Kumari *et al.*, 1975; Singh, 1977).

Change in phenolic compounds

Phenolic substances like phytoalexins are synthesized in the host cells in the presence of pathogen. They are non-specific toxic substances considered to play a role in disease resistance (Lakshman *et al.*, 1988). It has also been reported that phenolic materials in plant tissues may cause serious difficulties in the isolation and preservation of viruses (Matthews, 1970). Phenolic compounds do not appear to be responsible for necrotic hypersensitivity.

Delaying symptoms

The inoculations with biofertilizers either singly, dualy or all in combinations result in delay in appearance of the symptoms or clearly show low magnitude of disease symptoms (Deokar, 1998).

Production of growth promoting substances

Various compounds of the plant growth regulators have been found to reduce the virus content of plants or to mask the symptoms of diseases. These compounds include naphthalene acetic acid, indole butyric acid, gibberellic acid, kinetin, other cytokinins and related compounds (Daft, 1965; Aldwinckle, 1975). The growth promoting substances have been reported to be produced by *Azotobacter* and *Azospirillum* (Reynders and Vlassak, 1979). The most of the reports indicated that soil borne fungi could not inhibit the biofertilizer agents except that of Iswaran and Subba Rao (1966) who recorded that *Azotobacter chroococcum* is inhibited by soil fungi like *Cephalosporium*.

Role of Mycorrhizae in organic farming

Albert Bernard Frank (1885) defined the symbiotic relationship between fungi and plant roots as 'Mycorrhiza' (Greek meaning fungus root), attracting worldwide attention. Mycorrhizae are not pathogenic soil fungi that invade most absorbing roots of plants thereby forming a distinct morphological structure. Mycorrhizae (fungal roots) are the normal means by which all higher plants take up water and minerals from the soil. The mycorrhizal fungus parasitise the cortical cells of the root, thereby form a symbiotic relationship with the host. Read (1992) have proposed classification of mycorrhizae.. Recently, Sylvia (1999) summarized the characteristics of each type of mycorrhiza. Mycorrhizal associations of the fungi and plants can be classified into broad groups.

1. Ectomycorrhizae

Fungal symbiont septate (rarely aseptate), fungal penetration absent, fungal hyphae present between root cortical cells, producing the 'Hartig net', mantle or fungal sheath is usually present, sheath covering completely the absorbing root. The mantle can vary widely in colour, thickness and texture. Fungal symbiont usually belongs to Basidiomycotina (over 4000 fungal species are known to form ectomycorrhizal association) and only a few Ascomycotina members. Many genera belonging to Pinaceae (Gymnosperms), Betulaceae, Fagaceae, Myrtaceae, Dipterocarpaceae (angiosperm) are common components of temperate and sub-tropical forests.

2. VA-Mycorrhizae

Fungal symbiont aseptate and penetrates and develops highly branched arbuscules within the root cortical cells. Some arbuscular mycorrhizal (AM) fungi also produce vesicles, asexual spores and auxiliary cells and the fungal symbionts belong to Glomales (Morton, 1998). The term vesicular-arbuscular mycorrhizae (VAM) was used earlier to this type of endomycorrhiza, now it is clear that a few members (Gigasporiaeae) do not develop vesicles in roots; as sac. AM has been preferred acronym. VAM are seen in both herbaceous and woody chlorophyllous plants, e.g. Bryophytes, Pteridophytes, Gymnosperms and Angiosperms and very rarely in achlorophyllous plants. AMs are endomycorrhizal.

3. Ectendomycorrhizae

Fungal symbiont septate hyphae enter cells, fungal sheath present or rarely absent; hearting net and hyphal coils in cells present; haustoria absent; fungi belong to Basidiomycotina, autobionts chlorophyllous members of Angiosperms and Gymnosperms.

4. Ericaceous Mycorrhizae

Ericaceous mycorrhizal associations are found in the plants belonging to Ericales. These are endomycorrhizal as the fungal hyphae can penetrate the host cells. This group is divided into three major forms : (i) ericoid, (ii) arbutoid and (iii) monotropoid.

(i) Ericoid Mycorrhizae : Fungal partner septate hyphae enter cells. Cells of inner cortex become packed with hyphae; a true mantel and hartig net absent; a loose weft of hyphae develops over the root surface. Fungal symbiont belongs to the genus Hymenoscyphus (Ascomycetus). *Calluna rhododendron* and *Vaccinium* species. Typically growing in acid, peaty soils form ericoid mycorrhizal associations.

(ii) Arbutoid : Septate hyphae of the fungal symbiont enter host cells; fungal sheath and hartig net are present; hyphal coils present in cells, haustoria absent, host plant chlorophyllous; rarely achlorophyllous; fungal partner belongs to Basidiomycetes. Members of Ericales like *Arbutus, Arctostaphylos* and several species of *Pyrolaceae* develop arbutoid mycorrhizas.

(iii) Monotropoid : Fungal symbiont septate, fungal sheath and hartg net present, hyphae enter cells; hyphal coils present in cells; haustoria present; fungal partner belong to Basidiomycetes; host achlorophyllous members of Monotropaceae.

(iv) Orchid Mycorrhizae : The fungi involved in the symbiosis belong to Basidiomycetes; hyphae septate e.g. *Fomes, Marasmius* (common wood decaying fungi) and *Armillaria* and *Rhizoctonia* (common pathogens on forest trees). The fungi play an important role in the life cycle of orchids, as they are known to supply carbon and vitamins to the developing orchid embryo. Achlorophyllous orchid members depend on the fungal partner for carbon supply throughout their life. The fungus enters the host cells and forms hyphal coils within the cells. These hyphal coils are active for only a few days, after sometime they loose turgor and degenerate and the nutrient contents of the degenerated hyphae are absorbed by the orchid.

Effect of Ectomycorrhizae

Omon *et al.* (1995) observed that inoculation with EM fungi such as *Amanita* spp., *Laccaria iaccata, Russula* spp. *Boletus* spp. and *Scleroderma columnase* promoted the growth of root cutting of *Shorea leprosulas* comparedto control. They found increase in height growth 28-75 %, no. of leaves 25 - 50 %, leaf area 24 - 50 % and the highest mycorrhizal root percentage on root cuttings (32 - 67 %) inoculated with *L. laccata* on Ryps -sand (1 : 1). Mantel and hartig net were formed completely on root cutting inoculated with above species. *Scleroderma columnare* proved more effective in increasing nutrient uptake of N and P. *Laccaria laccata* was more effective in absorbing P and N nutrient (0.1 % and 1.09 %), respectively when compared to control.

Mycorrhizal fungi can make an important contribution to the success of vegetation by following mechanisms :

1. Enhancing establishment and growth of mycorrhizal plants by increasing nutrient uptake.

2. Maintaining diversity by boosting the ability of mycorrhizal plants to compete for resource.
3. Contributing to efficient recycling of nutrients and thus long-term stability.
4. Stabilizing the soil.

Ecto-mycorrhizae are being used now for biological control of root diseases. They proved necessary protection to plants : i) by acting as physical barrier and ii) also by harbouring high population of antagonize rhizosphere microflora.

The ecto-mycorrhizal fungi like *Lacatarious deciosus, Pisolithus tinctorius, Boletus luteus* and *Leucopaillus cerelis* were found to inhibit growth of many root pathogens. Marks (1971) record that short leaf pine disease caused by *Phytophthora cinnamoni* was absent in plants with ecto-mycorrhiza compared to 90-100 % infection on nonectomycorrhizal plants. *Laccaria laccata* is strongly suppressed *Fusarium oxysporum* under nursery condition (Pinus). Rangarajan (1989) inoculated *Casuarina equisetifolia* with *Frankia* sp., *Pisolithus tinctorius* and *Rhizopogon luteolus* and found that inoculum containing *Frankia* alone increased the biomass by 32 %, *Frankia* + *P. tinctorius* increased by 55 % and *Frankia* + *R. luteolus* by 68 %. The plants that received inoculum exhibited 100 % ectomycorrhizal colonization and nodulation. The shoot height, mycorrhizal development, root collar diameter and dry matter production of *Pinus patula* were increased when inoculated with *L. laccata* and *Thelephora terrestris.* Darusman *et al.* (1995) demonstrated that *Scleroderma columnae* produced the highest antibacterial antibiotics. Both *S. columnae* and *Rhizopogon* spp. exhibited antagonistic action against *Fusarium* spp. and *Rhizoctonia* spp.

The plant growth promoting fungus (PGPF) are abundant in the rhizosphere region, which produce various growth promoting substances that promote growth of roots, absorb sufficient nutrients and increase the activity of beneficial organisms in the rhizosphere. As a result, PGPF enhances the plant growth, control soil borne diseases and also induce systemic resistance to phytopathogen, The PGPF normally includes enclosed mycorrhiza and *Trichoderma* spp. The AMF naturally exist different cropping ecosystem and enrichment of their population in the rhizosphere profoundly increased plant growth in wide range of crops. Mycorrhiza is also used for the growth and establishment of tissue culture derived plants which will have beneficial effects like improved water relations , uptake of nutrients and tolerance to pathogenic infection, which in turn will result in enhanced growth, survival and establishment of plantlets.

REFERENCES

Aldwinckle, H.S. 1975. Studies on Growth Promoting substances produced by micro-organisms. Virology. 66 : 341-343.

Allen, M.F. 1991. The Ecology of Mycorrhizae. Cambridge Univ. Press, U.K.

Anonymous, 2002. Agricultural Research data book, Indian Agricultural Statistics Research institute Library Avenue, New Delhi, India pp. 73-77.

Anonymous, 2002-2003. Economic survey of India, pp. 188.

Anonymous, 2002-2003. Economics survey of Maharashtra. Directorate of Economics and Statistics, pp. 110-163.

Bhowmik, S.N. 1995. Investigations into isolation and characterization of *Acetobacter diazotrophicus* strains. M.Sc. (Agri.). Thesis. M.P.K.V., Rahuri.

Boddy, R.M., S. Urquiaga, V.M. Reis and J. Dobereiner 1991. Biological nitrogen fixation associated with sugarcane. Plant and Soil, 137 : 111-117.

Borkar, S.G.; V.K. Mandhare; K.N. Chaudhari; B.S. Patil and B. Varma. 1999. Studies on *Pseudomonas fluorescence* and *Trichoderma* spp. Indian J. Plant Pathol. 17 (1-2) : 78-81.

Cavalcante, V.A. and J. Dobereiner 1988. A new acid tolerant nitrogen fixing bacterium associated with sugarcane. Plant and Soil 108 : 23-31.

Cook, R.J. and A.D. Rovira. 1983. The nature and practices of biological control of plant pathogen. Am. Phytopathol. Soc.,scc. St.Paul.Min. pp: 533-540.

Daft, M.J. 1965. Studies on Growth Promoting substances produced by micro-organisms. Ann. Appl. Biol. 55 : 51-56.

Darusman, B.L., Y. Fakura, A.N. Anwar and M.D.S. joardjo. 1995. Studies on Growth Promoting substances produced by micro-organisums.Biol. Biotech. Mycor. Proc. II, Jakarta. Pp. 101.

Defago, G.; C.H. Berling; U. Burger, D. Hass, G. Kahr, K. Keel, C. Voisard, P. Wirthner and B. Wuthrick 1990. Suppression of black root rot of tobacco and other root diseases by strains of *Pseudomonas fluorescence.* Potential application and mechanism. In : Biological Control of Soil Borne Plant Pathogen. (Ed., Hornby, D). CAB International Wellingabord. Oxon, U.K. pp. 93-108.

Deokar, K.P. 1998. Studies on inhibition of cucumber mosaic virus in chilli by biofertilizers. M. Sc.(Agri.) Thesis submitted to Mahatma Phule Krishi Vidyapeeth, Rahuri. pp. 23-66.

Dubeikovsky, A.N., F.A. Mordukhova, V.V. Kochethov, F.Y. Polikarpova and A.M. Boronin. 1993. Plant Growth Promoting Rhizoctonia (PGPR) Soil Boil. Biochem. 25 : 1277-1281.

Frank, A.B. 1985. Symbiotic relationship between fungi and plant roots. Ber. Dt. Bot. Gos. 35 : 128-145.

Gaur, A. C. 1985. Phosphate solubilizing microorganisms and their role in plant growth and crop yield. Proc. Soil Biology Symposium, HAU, Hisar pp., 125-138.

Gaur, A. C. 1990. Phosphate solubilizing micro-organisms as biofertilizers. Omega publishers, New Delhi, pp. : 1-3.

Harley, J.L. and S.E. Smith. 1983. Mycorrhizal symbiosis. Academic Press, London.

Iswaran, V. and N.S. Subba Rao. 1966. Studies on *Azotobacter chroococcum* for inhibiting soil fungi. Indian Phytopath. 19 : 87 - 91.

King E.O., M.K. Ward and D.F. Rancy. 1958. Composition of Kings B medium for isolation of *Pseudomonas*. J. Lib. Clin. Med. 44 : 301-307.

Kloepper, J.W., J. Leng, M. Teintze and M.N. Schroth. 1980. Studies on fluorescent pigment with UV radiation. Carr. Microbiol. 4 : 317-320.

Lakshmanan, M.P. Gunasekaran R.N. Ubalthoose and P. Sundersen. 1988. Production of phytoalexin in cowpea infected with VA-mycorrhizal fungi. In : Mycorrhizal Round Table. (Eds. Varma, A.K., A.K. Okra, K.G. Mukerji and K.V.B.R. Tilak). J.N.K.V., Delhi. pp. 530 - 534.

Lakshmi-Kumari, M., K. Vijaya Lakshmi and N.S.S. Subba, Rao. 1975. Production of Antifungal compounds by *Azotobacter*. Phytopath. 75 : 27-30.

Lin, L., J.W. Kloepper and S. Tarun. 1995a. Studies on fluorescent *Pseudomonas* against fungal, bacterial and viral diseases. Phytopathology 85 : 695-698.

Lin, L., J.W. Kloepper and S. Tarun. 1995b. Studies on Fluorescent Pseudomonas against fungal,bacterial and viral diseases. Phytopathology 85 : 843-847.

Loper, J.E., 1988. Effect of Siderophores on damping off cotton. Phytopathology. 78 : 166-172.

M'piga, P., R.R. Belanger, T.C. Paulitz, and N. Benhamou. 1997. Studies on fluroscent *Pseudomonas* against infection of *Fusarium oxysporium* Physiol. Mol. Plant Pathol. 50 : 301-320.

Marks, D.H. 1980. Studies on Ectomycorrhizae root fungi .Trop. Myco. Res. (Ed., Mibolo, E.D.) pp. 13-71.

Marks, S.G.C. 1971. Comparison of Ectomycorrhiza with non-ectomycorrhizal plants. Ectomycorrhizae. Their Structure and Function. Acad. Press, New York.

Matthews, R.E.F. 1970. Role of phenolic compounds in isolation and preservation of viruses. Plant Virology. Acad. Press, New York. pp. 35-39.

Maurhofer, M., C. Reimmann, S.P. Sacherer, J. H., Heeb, J. Hase, and G. Defago. 1998. Biochemical and physiological changes to synthesis of defense proteins. Phytopathology. 88 : 678-684.

Maurhofer, M., P. Hase, J.P. Meuwly, P. Metraux and G. Defago. 1994. Correlation with Fluorescent bacteria and RP protein and PAL enzyme. Phytopathology 84 : 139-146.

Omon, R.M., S. Fakura and E. Suhendag. 1995. Studies on Ectomycorrhizal fungi. Proc. II A Comp. pp. 181.

Pannuri, R. 2002. Introduction of Mycorrhiza. In : Fundamentals of Mycorrhizal Biology and Biotechnology (Ed. Pannuri, R.). Agrotech Publ. Acad. Udaipur. pp. 7-16.

Purashothaman, D. and K. Vijila, 2000. Commercial production of *Acetobacter diazotrophicus* Biofertilizer. Biofertilizer newsletter Dec.pp. 9-13.

Ramamoorthy, V. and R. Sanutaooab 2001. Fluroscent *Pseudomonas* is a new strategy for plant disease management. J. Mycol. Pl. Path. 31(2) : 146-155.

Ramamoorthy, V., R. Vaishwanathan, T. Raghuchander, T. Prakasam and R. Samiyappan. 2001..Effect of . *Pseudomonas fluorescence* against *Colletotricum capsici.* Crop Protect. 20 : 1-11.

Rangarajan, M.D. Kandaswamy, G. Oblisami and R. Narayan. 1989. Studies of *Frankia* spp. Proc. I. Acom. pp. 288-291.

Read, D.J., D.H. Lewis, A.H., Fitter and I.J. Alexander. 1992. Mycorrhizas in Ecosystems. CAB International, Walling Ford, London.

Ramarethinam. S. and Chandra. K. (2006) Studies on the effect of Potash solubilizing / mobilizing bacteria *Frateuria aurentia* on Brinjal *Solanum melongena* L.) growth and yield. Pestology: 30(11) : 35-39.

Reynders, L. and K. Vlassak. 1979. Production of Growth promoting substances by *Azotobacter* and *Azospirillum.* Soil Biol. Biochem. 11 (1-6) : 547-548.

Sahoo. S. h. (2002) Potassium solubilizing bacteria and its prospects in different soil.: A review in National seminar on "Biotechnology"

Microbes to Man", School of Life Science Utkal University, Bhubaneshwar, Orissa pp. 93-95.

Sharma, G.D. and R.R. Mishra. 1986. Studies on mass inoculation of ectomycorrhizae . Acta Mycologica. 21 : 91-99.

Singh, T. 1977. Studies on interaction between *Azotobacter chroococcum* and some plant pathogens. I.A.R.I, Ph.D. Thesis, New Delhi.

Sneh. B.M., Y.E. Dupler and R. Baker, 1984. Effect of Siderophore on vascular wilts caused by *Fusarium oxysporium*. Phytopathology. 74 : 1115-1124.

Stutz, E., G. Defago and H. Kern. 1986. Studies on *Pseudomonas fluorescence* to control tobacco black rot. Phytopathology. 76 : 181-185.

Subba Rao, N.S. 1988. Biofertilizers in Agriculture, Second edition, Oxford and IBH Publishing Co. Pvt. Ltd. pp. 208.

Subba Rao, N.S. 1993. Biofertilizers in Agriculture and Forestry. Third revised edition. Oxford and IBH Publ. Co. Pvt. Ltd., New Delhi pp. 129-131.

Thaper, H.S. 1988. Mycorrhiza Round Table. Proc. Nat. Workshop. pp. 33-52.

Thomashow, I.S. and D.M. Weller. 1996. Current concepts in the use of introduced bacteria for biological disease control. Mechanism and antifungal metabolite. In : Plant Microbe Interaction. Vol. I. (Ed, Staccy G.K.). Chapman and Hall. New York, pp. 187-235.

Tilak, K.V.B.R. 1993. Studies on Nitrogen fixing biofertilizers. Bacterial fertilizers. ICAR, New Delhi.

Varshney, S., H.S. Chaube and H.S. Singh. 2000. Antagonistic Potential against plant pathogens. Indian J. Plant Pathol. 18 (1-2) : 40 - 43.

Vishwanathan, R. and R. Samiyappan, 1990. Biochemical and physiological changes to synthesis of defense protein. Sugar. Tech. 1 : 67-76.

Wani, P.V. 1980. Studies on phosphate solubilizing micro-organisms. A review. J. Maharashtra agric. Univ. 5(2) : 114-147.

Wani, P.V. and Konde, B.K. 1986. Role of *Azotobacter* and Vesicular Arbuscular Mycorrhizae J. Maharashtra agric. Univ. 11(3) : 263-268.

Wei, L., J.W. Kloepper and S. Tarun. 1996. Fluorescent *Pseudomonas* as seed dresser and soil application. Phytopathology 86 : 221 - 224.

Xu, G.W. and D.C. Cross, 1986. Effect of Siderophore on potato decay caused by *Erwinia caratovora*. Phytopathology. 76 : 414 - 421.

❑❑❑

3

Use of Microalgae and Cyanobacteria as Biofertilizers and Soil Conditioners

J.B. Khan, Manoj Kumar and G.P. Singh

INTRODUCTION

India is rich in the possession of large number of water bodies, fresh water as well as saline ones. They have great diversity of phycoflora. A large number of algal species are found in soil also. Algae found in water as well as in soil are useful for various purpose and they play a role to maintain soil fertility. Cyanobacteria inhabit nearly all illuminated environments on Earth and play key roles in the carbon and nitrogen cycle of the biosphere. They are found in almost every possible habitat, from salt water to fresh water or bare rock to soil. They are also found in extreme conditions of temperature, for example in the thermal springs or in the earth's crust in the desert. Some of the reasons for the success of cyanobacteria in modern habitats can be related to their long evolutionary history. Symbiotic interactions between cyanobacteria and other organisms are amazingly diverse, including examples from many of the major phyla of plants and animals. Some cyanobacteria belong to the few groups of organisms that can convert inert atmospheric nitrogen into an organic form, such as nitrate, ammonia or urea. Plants need for their growth this organic form of nitrogen, which must be obtained from soil. Therefore, cyanobacteria are very important organisms for the health and growth of many plants.

Cyanobacteria are employed in agriculture as biofertilizers and soil conditioners. The majorities of cyanobacteria are capable of fixing atmospheric nitrogen and are effectively used as biofertilizers (Venkataraman, 1993; Anand, 1998; Vaishampayan *et.al.*, 2001). Algalization benefit crop plants through excreting part of the biologically fixed nitrogen, secreting growth promoting substances and different types of secondary metabolites; adding appreciable amounts of organic matter into the soil, solubilizing insoluble phosphates and improving the physical and chemical nature of the soil (Goyal, 1993).

NITROGEN FIXATION

Nitrogen fixation generally refers to the natural process, either biological or abiotic, by which nitrogen in the atmosphere is converted into ammonia. This process is essential for life because fixed nitrogen is required to biosynthesize the basic building blocks of life. Nitrogen is an essential constituent of macromolecules such as proteins, nucleic acids, chlorophylls, enzymes and other physiological substances in green plants. Nitrogen is one of the macronutrient that is required in high amounts by plants, and its availability in the soil may change substantially in relatively short time intervals (Cameron and Haynes, 1986). For rapid growth of all plants, nitrogen is the most common limiting factor. Hence, an adequate supply of nitrogen in agriculture is very important factor (Chuang, 1984).

The species of cyanobacteria which are known to fix atmospheric nitrogen are classified into three groups (i) Heterocystous-aerobic forms, (ii) Aerobic unicellular forms and (iii) Non-heterocystous, filamentous, microaerophilic forms. Many cyanobacteria have the capacity to manufacture nitrogenase. Because the enzyme complex is anaerobic, significant fixation by unicellular, colonial and some filamentous species occurs only in the absence of air. Therefore, only heterocystous species are valuable as biofertilizers. The machinery for fixing nitrogen has been shown to be present in all the currently recognized taxonomic groups of cyanobacteria. A phylogenetically related group of filamentous cyanobacteria, that includes genera such as *Anabaena, Nostoc,* and *Fischerella* of the taxonomic classification of Rippka (Rippka *et.al.*, 1979) has developed a complex and efficient mechanism for performing nitrogen fixation under aerobic conditions. This consists of the development of heterocyst, specialized cells that, under conditions of aerobiosis and combined nitrogen deprivation, differentiate from vegetative cells at semi-regular intervals in the filament and bear a series of modifications devoted to the protection of nitrogen fixation apparatus from oxygen. During the nitrogen fixation reaction, enzyme nitrogenase catalyses the ATP-dependent reduction of nitrogen to two molecules of ammonium. A region of the chromosome of the heterocyst-forming cyanobacterium *Anabaena* sp. containing the *nifHDK* genes, encoding the three polypeptides of nitrogenase, was identified and cloned by means of hybridization with the *Klebsiella pneumoniae nif* genes (Mazur *et.al.*, 1980; Rice *et.al.*, 1982).

NITRATE ASSIMILATION BY CYANOBACTERIA

For cyanobacterial nutrition nitrate is almost certainly the most abundant source of combined nitrogen. The assimilation of nitrate by cyanobacteria involves nitrate uptake and reduction of intracellular nitrate (via nitrite) to ammonium, which is the nitrogen form incorporated into organic compounds. Nitrite, which can also fulfill the nitrogen requirement of cyanobacteria, is taken up into the cell and then reduced to the level of ammonium. Intracellular nitrate is reduced to nitrite in a two-electron reaction catalyzed by nitrate reductase (Nar), the resulting nitrite being then reduced to ammonium in a six-electron reaction catalyzed by nitrite reductase (Nir). Both Nar and Nir use reduced feredoxin as physiological electron donor (Arizmendi and Serra, 1990; Manzano *et.al.*, 1976; Mendez *et.al.*, 1981).

NITROGEN ASSIMILATION IN CYANOBACTERIA

The cyanobacteria mainly use inorganic compounds (dinitrogen) to fulfill their nitrogen requirements, but urea and other organic sources of nitrogen, such as some amino acids, nitrate and ammonium, can also be assimilated by some cyanobacteria. Nitrogen can represent as much as about 11% of the dry weight of a cyanobacteria (Wolk, 1973).

The urea and inorganic nitrogen sources other than ammonium are first metabolized to ammonium to allow assimilation of their nitrogen atoms. Some amino acids like arginine are assimilated by the production of ammonium among their metabolic products. The cyanobacteria played central position in nitrogen metabolism by the glutamine synthetase/glutamate synthase (GS/GOGAT) cycle for ammonium assimilation. Glutamate produced in this pathway is not only the major nitrogen donor for the biosynthesis of other nitrogen containing metabolites but is itself a precursor for some amino acids and 5-aminolevulinate, the immediate precursor for porphyrin, phycobilin and chlorophyll synthesis. Glutamine also donates nitrogen to some metabolites. The amino acids arginine and aspartate together make up cyanophycin, a unique storage reservoir of nitrogen found in many cyanobacteria.

MICROALGAE AS SOIL CONDITIONERS

Soil conditioners

Soil conditioners are the products which are added to soil to improve its quality. They can be used to rebuild soils which have been damaged by improper management, to make poor soil more usable, and to maintain soil in peak condition. Many studies have been conducted in proving the productivity of sandy soil using synthetic and natural soil conditioners. Conditioners increase water use efficiency by reducing the evaporation rates and raising the water holding capacity in coarse textured soil. Over the past two decade, many studies have evaluated the effects of soil conditioners on the physical and hydrological properties of soil. These studies has shown that soil conditioners when applied to coarse textured soils improve and stabilize soil aggregation (Al-Omran *et.al.*, 1987; Al-Omran and Shalaby, 1992), raising water holding capacity (Choudhary *et.al.*, 1995), control soil erosion (Levy *et.al.*, 1992) and suppress water evaporation from soil (Choudhary *et.al.*, 1995). Soil conditioners can reduce evaporation and increase the amount of water conserved (Johnson, 1984; Choudhary *et.al.*, 1995). There are many types of natural conditioners which recover the physical properties of soil and increase its fertility. Their use decreases the crop production cost and reduces environmental pollution. The most commonly used natural conditioners are organic manure and microalgae (Metting, 1990; Mandal *et.al.*, 1999).

Several investigators reported the significant role of soil conditioners on plant growth and its yield. The yield of tomato plants grown in yellow podzolic soil in the green house was increased with the addition of various types of soil conditioners (Sjamsudin *et.al.*, 1994).

MICROALGAE AND CYANOBACTERIA AS SOIL CONDITIONERS

In sprinkler-irrigated farmlands eukaryotic, unicellular, green microalgae were applied as soil-conditioning agents for soil erosion control in the temperate zones. These technologies have great potential, but they suffer variable success because of the absence of quality inoculants and poor understanding of soil microbial ecology. Concerted isolation, screening and strain improvement programs together with superior product formulation technologies that hasten efforts to realize the potential of microalgae application in agriculture (Metting, 1990). Earlier studies reported that cyanobacteria improve soil structure by increasing soil aggregation, water holding capacity and soil aeration; therefore, its application is useful for the reclamation of soils (Rogers and Burns, 1994; De-Caire *et.al.*, 1997; Hegde *et.al.*, 1999). Rai *et.al.*(2000) reported that cyanobacteria are good colonizers of the nitrogen poor soil, and through their nitrogen input into the environment they may help to create habitats suitable for other species.

The effect of cyanobacteria *Nostoc* strains on two types of clay soil from Tuscany were studied in order to show not only aggregation initiation but also protection of soil porosity due to reduction of the damaging effect of water addition (Falchini *et.al*, 1996). Aggregating effects of cyanobacteria were determined on two soil types, *viz.*, garden soil and artificial soil mixture composed of three parts of garden soil and one part of sand. The percentage increases in soil aggregation varied from 35.7% to 42.8% in the garden soil and from 57.8% to 78% in the sandy soil mixture. The results indicate that these cyanobacteria brought about considerable aggregation of loose unbound soil particles. These algae play—significant role in improving the physical quality of the soil and thus help in checking soil erosion to a certain extent (Marathe, 1972).

The potential of the nitrogen fixing cyanobacteria *Nostoc muscorum* in improving the aggregate stability of a poorly structured silt loam soil was studied in the green house by Rogers and Burns (1994), aggregate stability increased by an average of 18%. Inoculation with *N. muscorum* also had a pronounced effect on soil chemical and biological properties. Erosion causes loss of soil fertility and affects vegetational succession. The algae prevent soil erosion by binding particles of soil. Sand and clay particles are immobilized by a dense network of filaments of the two dominating cyanobacterial species, *Microcoleus vaginatus* and *M. chthonoplastes*, which secrete sheaths to which particles adhere. These microorganisms can tolerate long periods of desiccation and are capable of instant reactivation and migration following wetting (Campbell, 1979). Vazquez *et.al.*(1997) suggested that algae maintain substrate humidity and favors germination of the studied species. This could be important in sand dune slacks where marked fluctuations in temperature and humidity occurred.

Cyanobacteria can fix about 25 kg nitrogen/ha/season. Apart from nitrogen fixation, inoculation with cyanobacteria is also reported to reduce considerably the total sulphides and ferrous iron content of the soil (Aiyer, 1965). The cyanobacterial ability to mobilize insoluble forms of inorganic phosphates is evident from the finding of Kleiner and Harper (1977) who reported more extractable phosphates in soil with cyanobacterial cover than in nearby soil without cover.

BIOFERTILIZERS

One of the major concerns in today's world is the pollution and contamination of soil. The use of chemical fertilizers and pesticides has caused tremendous harm to the environment. An answer to this is the biofertilizer, an environmentally friendly fertilizer now used in most countries. Biofertilizers are ready to use live formulates of such beneficial microorganisms which on application to seed, root or soil mobilize the availability of nutrients by their biological activity in particular, and help build up the micro-flora and in turn the soil health in general. These are the products containing microorganisms which have the ability to mobilize nutritionally important elements from a non-usable to a usable form through biological processes. Microorganisms play an important role in various chemical transformations of soils and thus, influence the availability of major nutrients like nitrogen, potassium, phosphorus and sulphur to the plants. Cyanobacteria and phosphate solubilizing bacteria were used as biofertilizers to increase crop production (Singh, *et.al.*,1997; Earanna and Govindan, 2002).

In Indian agriculture, biofertilizers have emerged as promising components of the integrated nutrient supply system in. Among biofertilizers benefiting the cereal crop production are *Azotobacter, Azospirillum, Azolla,* cyanobacteria, P-solublising microorganisms and mycorrhizae. The beneficial properties of biofertilizers are addition of nitrogen, increased soil organic matter and soil aggregation (Kulik, 1995). Synergistic effects with beneficial microorganisms, excretion of organic acids that increase phosphorus availability, decrease of sulphide injury by increased O_2 content (Ordog, 1999).

Chemical fertilizers are expensive, they disturb the equilibrium of agro-ecosystems and cause pollution to the environment. These problems may be avoided by the use of biofertilizers. The role of biofertilizers in sustainable agriculture recorded special significance, particularly in the present context of high cost of chemical fertilizers. The production and application of biofertilizers to leguminous plants, millets, oil seeds, rice, and forest nursery plants are very common in India (Kannaiyan, 2002; Rai *et.al.*, 2003).

ROLE OF MICROALGAE AND CYANOBACTERIA AS BIOFERTILIZERS AND NITROGEN FIXERS

Cyanobacteria and microalgae were used as biofertilizers in the coastal regions of all over the world from ancient times. The rational background for this interesting utilization of microalgae or their extracted residues is the increase in water-binding capacity and mineral composition of the soil (Critchley and Ohno, 1998). The important role of microalgae in the soil ecosystem has often been neglected. The beneficial effects originate not only from the production of compounds for particle adherence and water storage in soils or nitrogen-fixing, but also from bioactive compounds produced by algae which influence higher plants (Borowitzka, 1995; Metting, 1996). Soil microalgae should be regarded by microalgal biotechnologists as a promising area to find new species with unexpected properties. While nitrogen fixation with microalgae (*Anabaena, Nostoc*) is important for rice production in tropical and subtropical agriculture, surface solidification against erosion processes is also of interest in more arid regions. During the past decade, plant growth regulators from both macro- and microalgae gained increasing attention. Substances or extracts were found which promote germination, leaf or stem growth or flowering.

Heterocystous cyanobacteria and several non-heterocystous cyanobacteria are known for their ability to fix atmospheric nitrogen. The fertility of many tropical rice field soils has been mainly attributed to the activity of nitrogen fixing cyanobacteria. Inoculation of cyanobacteria to increase the fertility of soils has been successfully attempted. For example, *Azolla* was used as an organic fertilizer in rice cultivation in many countries (Kaushik and Venkataraman, 1979). Addition of *Azolla* was found to support the growth of soil micro-organisms including heterotrophic nitrogen fixers. Recently, nitrogen fixing cyanobacteria have been reported to dominate desert crusts worldwide (Garcia-Pichel and Pringault, 2001). This is believed to contribute significantly to the fertility of desert soils and may eventually facilitate vegetation of deserts.

Mian (1984) indicated that the nitrogen produced by free living cyanobacteria is immediately available to crops. A wide range of nitrogen fixing cyanobacteria exists in rice field ecosystems (Watanabe and Roger, 1984; Watanabe *et. al.*, 1987). Nitrogen fixed by the symbiotic association of cyanobacteria (cyanobionts) is transferred and used by various plant groups other than rice (Stewart *et.al.*, 1983; Rai, 1990; Kannaiyan *et.al.*,1997). *Anabaena* and *Nostoc* have been recorded among the common nitrogen fixing cyanobacteria in rice fields. Similarly, more than hundred strains of heterocystous cyanobacteria belonging to the genera *Anabaena, Nostoc, Nodularia, Cylindrospermum, Scytonema, Calothrix, Anabaenopsis, Mastigocladus, Chlorogloeopsis, Fischerella, Nostochopsis, Rivularia, Tolypothrix, Aulosira, Stigonema, Hapalosiphon, Cauptylonema, Gloeotrichia, Schytonematopsis, Westiella, Westiellopsis, Wollea* and *Chlorogloea* have been found to be efficient as nitrogen fixers (Venkataraman, 1993). The nitrogen fixing capacity of the cyanobacteria species was found to be correlated with their biological activities. Twenty five species of cyanobacteria were isolated from the natural habitat and from the enriched culture of paddy fields. Wide variations were observed in the growth rate and nitrogen fixing capacity among the species under each genus and between genera (Sahu, 2000). Many workers reported the ability of cyanobacteria to liberate high amounts of their nitrogenous substances into the medium. A large proportion of the extracellular nitrogenous materials are in the form of polypeptides and only a small fraction is in the form of free amino acids (Whitton, 1965).

The utilization of cyanobacteria as a biofertilizer is very promising as rice fields provided an ideal and favourable environment for proper growth of cyanobacteria. (Kerni, 1991; Gupta and Pandher, 1996; Ghosh and Saha, 1997; Kannaiyan, 2002). Nitrogen fixing cyanobacteria are the dominant microflora in rice fields and are currently used as a supplement to chemical nitrogen fertilizers for rice cultivation in rice-growing countries, including India and Bangladesh. This technology suffers from serious drawbacks and its use at farm level is not gaining universal acceptance due to some major problems. In order to significantly improve the efficient use of cyanobacteria as nitrogen based biofertilizer for rice cultivation experiments were carried out in different dimensions both in the laboratory and in the field. Cyanobacterial strains were isolated, identified and quantified from a wide range of distinctively different types of soil, *viz.* acid, calcareous, saline, red and neutral soil. These biofertilizers were recommended to be used in improving the soil environment. Increase of grain yield by over 25% was obtained in plots algalized with fresh cyanobacteria inoculums containing regional isolates in comparison with the yield of plots inoculated with soil based culture (Adhikary, 1999).

Cyanobacterial collections have been used to screen their growth and nitrogen fixing abilities in both sterile and non-sterile soil under different incubation periods. The effect of cyanobacterial exudates on germination of rice grains was also investigated. The most potent strains for nitrogen fixation in sterile soil were *Anabaena variabilis, Nostoc spongiaeforme* and *N. linckia* recorded after 90 days of incubation. *N. muscorum* and *Calothrix brevissima* recorded highest values of nitrogen fixation in the non-sterile soil at the same incubation period. Pretreatment of rice grains with cyanobacterial exudates of *N. linckia, N. muscorum* and *A. flos-aquae* stimulated germination by 25, 14 and 10 %, respectively. The results revealed that the nitrogen-fixing ability of the cyanobacterial exudates on rice grain germination appeared to be strain-specific (Omar, 2001). The *Azolla-Anabaena* symbiosis is outstanding due to its high productivity combined with its ability in fixing nitrogen at high rates. Because of this, in recent decades, countless studies have been conducted in this association, but with insufficient coordination (Kannaiyan *et.al.*, 1997; Wagner, 1997; Satapathy, 1998; Pabby *et.al.*, 2003). Inoculation of plants with non-symbiotic rhizosphere microorganisms for crop improvement has become a center of interest.

In recent years, there is a great interest in establishing novel associations between higher plants and various nitrogen fixing microorganisms. Cyanobacteria isolated from different soil types when inoculated into plant species of agricultural importance (wheat, corn, bean, sugar beet and rice), different types of associations between cyanobacteria and plants were observed, including penetration of the cyanobacteria into roots (Svircev *et.al.*, 1997).

Cyanobacteria biofertilizers have several advantages over chemical fertilizers. They are non-polluting, in-expensive, utilize renewable resources. In addition to their ability of using free available solar energy, atmospheric nitrogen and water. Beside supplying nitrogen to crops, they also supply other nutrients such as vitamins and growth substances (Wagner, 1997).

SECRETION OF OTHER ORGANIC COMPOUNDS BY MICROALGAE AND CYANOBACTERIA

Microalgae and cyanobacteria are beneficial for crop plants through excreting part of the biologically fixed nitrogen and secreting growth promoting substances like vitamins and amino acids, adding appreciable amounts of organic matter into the soil. These algae are mainly represented by green and blue-green species which are an additional source of organic substances in the soil (Kislitsina *et.al.*, 1978). Most of the cyanobacteria can produce exo-polysaccharides, amino acids and peptides in the algal filtrate (Ahmed *et.al.*, 2000; Kerby *et.al.*, 1989). In addition, several species were shown to release sugars and high molecular weight vitamin-B12 binding substances (Fogg, 1971). Several reports indicate an increase in organic matter and nitrogen content of soils inoculated with cyanobacteria under pot and field conditions (Singh and Singh, 1989; Vaishampayan, 1998). Inoculation of soil with *Nostoc muscorum* caused increase in total carbon and nitrogen, in addition to an increase in the soil indigenous microbial population. The increase in soil polysaccharides was 2.96-3.49 times in the soils inoculated with *Nostoc muscorum* in comparison with non-inoculated soils (Rogers and Burns, 1994). Rodgers *et.al.* (1979) reported that the filamentous cyanobacteria produce compounds that stimulate plant growth included *Anabaena sp.*, *Nostoc sp.* and *Nodularia sp.* Considerable amounts of free aminoacids were found in the cyanobacterial filtrates (Stewart, 1963; and Whitton, 1965).

Many cyanobacteria are known to produce different kinds of secondary metabolites such as auxins (Stewart, 1974; Serdyuk *et.al.*, 1991), auxin like substances (Venkataraman, 1981; Kaushik and Misra, 1989). They also have plant hormone-like activities and thus they influence growth of rice through the release of these metabolites (Bradley, 1991).

Microalgae and cyanobacteria are potential source of biologically active secondary metabolites. Investigations over the last decades have identified compounds with cytotoxic, antifungal, antibacterial or antiviral activities. Most of the isolated substances belong to groups of amides, alkaloids and peptides (Borowitzka, 1995; Falch, 1996; Kleinkauf and Von Dohren, 1997; Kreitlow *et.al.*, 1998).

Biofertilizers and soil conditioners produced by microalgae and cyanobacteria have great economic potential and they have ecological applications in the sense of waste water treatment and in the agriculture. They maintain soil fertility by nitrogen and other organic componds. Future trend seems to be the use of the biological activity of microalgal products against plant diseases caused by viruses or bacteria. It is likely that microalgae can be a source of a new class of biological plant protecting substances. Cyanobacterial utilization is limited due to the lack of basic knowledge regarding the factors involved in the success and failure of establishment of inoculated species. Little attention has been paid to the inoculums quality specially for testing the composition and viability of cyanobacteria in the soil based inoculants.

REFERENCES

Adhikary, S.P. 1999. Potentiality of cyanobacteria biofertilizer containing local isolates on the growth and yield of two different rice varieties. *Proc. Nat. Symp. Warangof.* British Library.

Ahmed, D., El-Gamal, M.S., Ammar, U.M. and El-Raouf, T.M. 2000. Role of some cyanobacteria in enhancement of soil characteristics. In: *Int. Symp. Phycol., Egypt. J. Plycol.* 1:99-106.

Aiyer, R.S. 1965. Comparative algological studies in rice fields in kerala state. *J. Agric. Res. Kerala.* 3:100-106.

Al-Omran, A.M., Mustafa, M.A., and Shalaby, A.A. 1987. Intermittent evaporation from soil columns as affected by gel-forming conditioners. *Soil Sci. Soc. Am. J.* 51:1593-1599.

Al-Omran. A.M., and Shalaby, A.A. 1992. Effect of water quality and gel-conditioner rate on intermittent evaporation. *J. King Saud Univ. Agri. Sci.* 4:273-286.

Anand, N. 1998. Blue-green algae (cyanobacteria) as biofertilizers : Retrospects and prospects. In: *Microbes: For health, wealth a sustainable environment.* (ed.) A. Varma. Malhotra Publication, New Delhi. pp. 65-71.

Arizmendi, J.M. and Serra, J.L. 1990. Composition of the ferredoxin- nitrite reductase from the cyanobacterium *Phormidium laminosum. Biochem. Soc. Trans.* 18: 637-638.

Borowitzka, M.A. 1995. Microalgae as sources of pharmaceuticals and other biologically active compounds. *J. Appl. Phycol.* 7: 3-15.

Bradley, P.M. 1991. Plant hormones do have a role in controlling growth and development of algae. *J. Phycol.* 27:317-321.

Cameron, K.C. and Haynes, R.J. 1986. Retention and movement of nitrogen in soils. In: *Mineral Nitrogen in the Plant-Soil System.* (ed.) R. J. Haynes. Academic Press, Orlando. pp.166-241.

Campbell, S.E. 1979. Soil stabilization by a prokaryotic desert crust : Implications for precambrian land biota. *Orig. life.* 9 (4):335-348.

Choudhary, M.I., Shalaby, A.A. and Al-Omran, A.M. 1995. Water holding capacity and evaporation of calcareous soils as affected by four synthetic polymers. *Commun. Soil Sci. Plant Anal.* 26:2205-2215.

Chuang, Y.T. 1984. Studies on the cultivation of nitrogen fixing microorganisms - Cultivation conditions of the blue green algae and preliminary application. *Plant Soils.* 31(3):84-87.

Critchley T, Ohno, M. 1998. Microalgae as source of pharmaceuticals and other biologically active compounds. *J. Appl. Algol.* 7:3-15.

De-Caire, G.Z., De-Cano, M.S., De-Mule, M.C., Palma, R.M. and Colombo, K. 1997. Exopolysaccharide of *Nostoc muscorum* (cyanobacteria) in the aggregation of soil particles. *J. Appl. Phycol.* 4:249-253.

Earanna, N. and Govindan, R. 2002. Role of biofertilizers in mulberry production - A review. *Indian J. Seric.* 41(2):92-99.

Falch, B. 1996. Was steckt in cyanobakterian. *Pharmazie in unserer zeit.* 25: 311-321.

Falchini, L., Sparvoli, E. and Tomaselli, L. 1996. Effect of *Nostoc* (cyanobacteria) inoculation on the structure and stability of clay soils. *Biol. Fertility Soils.* 23:246-252.

Fogg G.E. 1971. Extracellular products of algae in fresh water. *Hydrobiol.* 5:1-25.

Garcia-Pichel, F. and Pringault, O. 2001. Cyanobacteria track water in desert soils. *Nature.* 413:380–381.

Ghosh, T.K. and Saha, K.C. 1997. Effects of inoculation of cyanobacteria on nitrogen status and nutrition of rice (*Oryza sativa* L.) in an Entisol amended with chemical and organic sources of nitrogen. *Biology and Fertility of Soils. 24* (1):123-128.

Goyal, S.K. 1993. Algal biofertilizer for vital soil and free nitrogen. *Proc. Indian N. Sci. Acad.* 59(3-4):295-301.

Gupta, R.P. and Pandher, M.S. 1996. Biofertilizers in agriculture. *J. Res. Punjab Agric. Univ.* 33(1-4):209-224.

Hegde, D.M., Dwivedi, B.S. and Babu, S.N.S. 1999. Biofertilizers for cereal production in India- A review. *Ind. J. of Agric. Sci.* 69(2):73-83.

Johnson, M.S. 1984. Effect of soluble salts on water absorption by gel-forming soil conditioners. *J. Sci. Food Agric.* 35:1063-1066.

Kannaiyan, S. 2002. Biofertilizers for sustainable crop production. *Biotechnology of Biofertilizers.* 9-49.

Kannaiyan, S., Aruna, S.J., Kumari, S.M.P. and Hall. D.O. 1997. Immobilized cyanobacteria as a biofertilizer for rice crops. *J. Applied phycology.* 9:167-174.

Kaushik, B.D. and Misra, S. 1989. Growth promoting substances of cyanobacteria. Detections of amino acids, sugars and auxins. In: Proceedings of the Indian National Science Academy. Part B, *Biological Sciences.* 55(5-6): 499-504.

Kaushik, B.D. and Venkataraman, G.S. 1979. Effect of algal inoculation on the yield and vitamin C content of two varieties of tomato. *Plant Soil.* 52, 135–137.

Kerby, N.W., Rowell, P. and Stewart, W.D.P. 1989. The transport assimilation and production of nitrogenous compounds by cyanobacteria and microalgae. In: *Algal and Cyanobacterial Biotechnology.* (eds.) R.C. Cvesswell, T.A.V. Rees and N. Shah. Longman Scientific. Technical, Harlow. pp.50-90.

Kerni, P.N. 1991. Nitrogen biofertilizers in rice: seed, seedling and soil application. *Research and Development Reporter.* 8(1):60-64.

Kislitsina, V.P., Zhdanova, E.M. and Sudakova, E.A. 1978. Effect of crops on the humus accumulation process in the grey forest soils of Priangarye. *Zentra Lb1 Bakteriol Naturwiss.* 133(4):283-285.

Kleiner, K.T. and Harper, K.T. 1977. Soil properties in relation to cryptogamic ground cover in Canyon-lands National park. *J. Range Managem.* 30:202-205.

Kleinkauf, H. and Von, D.H. 1997. Products of secondary metabolism. In: *Biotechnology* (Eds.) H. J. Rehm and G. Reed. Weinheim. pp.308-309.

Kreitlow, S., Mundt, S. and Lindequist, V. 1998. Cyanobacteria - A potential source of new biologically active substances. *J. Biotechnology.* 70:61-63.

Kulik, M.M. 1995. The Potential for using cyanobacteria (bluegreen algae) and algae in the biological control of plant pathogenic bacteria and fungi. *European. J. Plant pathology.* 101(6):585-599.

Levy, G.J., Levin, J., Gal, M., Ben-Hur, M. and Shainberg, I. 1992. Polymers effects on infiltration and soil erosion during consecutive simulated sprinkler irrigation. *Soil Sci. Soc. Amer. J.* 56:902-907.

Mandal, B., Vlek, P.L.G. and Mandal, L.N. 1999. Beneficial effects of blue-green algae and *Azolla*, excluding supplying nitrogen, on wetland rice fields : A review. *Biol. Fertil. Soils.* 28:329-342.

Manzano, C., Candau, P., Gomez-Moreno, C., Relimpio, A.M. and Losada, M. 1976. Ferredoxin-dependent photosynthetic reduction of nitrate and nitrite by particles of *Anacystis nidulans*. *Mol. Cell Biochem.* 10:161-169.

Marathe, K.V. 1972. Role of some blue-green algae in soil aggregation. In: *Int. Symp. on Taxonomy and Biology of Bluegreen algae,* 328-331.

Mazur, B.J., Rice, D. and Hasekom, R. 1980. Identification of blue-green algal nitrogen fixation genes by using heterologous DNA hybridization probes. *Proc. Natl. Acad. Sci. USA.* 77:186-190.

Mendez, J.M., Herrero, A. and Vega, J.M. 1981. Characterization and catalytic properties of nitrite reductase from *Anabaena sp.* 7119. *Z. Pflanzenphysiol.* 103:305-315.

Metting, B. 1990. Microalgae applications in agriculture. *Dev. Ind. Microbiol.* 31:265-270.

Metting, F.B. 1996. Biodiversity and application of microalgae. *J. Ind. Microbiol.* 17:477–489.

Mian, M.H. 1984. A 15N tracer study to differentiate nitrogen supplied to flooded rice plants by *Azolla* and *Anabaena* during their early and later stages of decomposition. *Ind. J. Agric. Sci.* 54:733-800.

Omar, H.H. 2001. Nitrogen-fixing abilities of some cyanobacteria in sandy loam soil and exudates efficiency on rice grain germination. *Bull. Fac. Sci. Assiut. Univ.* 30(1- D):111-121.

Ordog, V. 1999. Beneficial effects of microalgae and cyanobacteria in plant soil-system with special regard to their auxin-and cytokinin-like activity.In: *International workshop and training course on microalgal biology and biotechnology,* Moson-magyarovar, Hungary. pp.43-44.

Pabby, A., Prasanna, R. and Singh, P.K. 2003. *Azolla- Anabaena* symbiosis-from traditional agriculture to biotechnology. *Ind. J. Biotech.* 2:26-37.

Rai, A.N. 1990. *CRC Hand Book of Symbiotic Cyanobacteria.* CRC Press, Boca Raton.

Rai, A.N., Sodeback, E. and Bergman, B. 2000. Cyanobacterium-plant symbiosis. *New Phytologist.* 147:449-481.

Rai, U.N., Pandey, K., Sinha, S., Singh, A., Saxena, R. and Gupta, D.K. 2003. Revegetating fly ash landfills with *Prosopis juliflora* L. : Impact of different amendments and *Rhizobium* inoculation. *Environment International.* 30:293-300.

Rice, D., Mazur, B.J. and Haselkorn, R. 1982. Isolation and physical mapping of nitrogen fixation genes from the cyanobacterium *Anabaena* 7120. *J. Biol. Chem.* 257:13157-13163.

Rippka, R., Deruelles, J., Waterbury, J.B., Herdman, H. and Stanier, R.Y. 1979. Generic assignments, strain histories and properties of pure cultures of cyanobacteria. *J. Gen. Microbiol.* 111:1-61.

Rodgers, G.A., Bergman, B., Henriksson, E. and Uridis, M. 1979. Utilization of blue green algae as biofertilizers. *Plant soil.* 52:99-107.

Rogers, S.L. and Burns, R.G. 1994. Changes in aggregate stability, nutrient status, indigenous microbial populations, and seedling emergence, following inoculation of soil with *Nostoc muscorum. Biology and Fertility of Soils.* 18(3):209-215.

Sahu, J.K. 2000. Growth and nitrogen fixation of cyanobacteria from rice fields of Puri district, Orissa. *Ad. Plant Sci.* 13(1):47-50.

Satapathy, K.B. 1998. Comparative efficiency of blue-green algae, *Azolla* and other biofertilizers on growth of rice. *Indian. J. Plant physiol.* 4(2):100-104.

Serdyuk, O.P., Smolygina, L.P., Kobzar, E.V. and Gogotov, I.N. 1991. Exometabolites with phytohormonal biological activity formed by an *Azolla-Anabaena azollae* symbiotic association. *Doklady. Biochem.* 316(1-6):8-10.

Singh, A.L. and Singh, P.K. 1989. Phosphorus fertilization and the growth and N2-fixation of *Azolla* and blue-green algae in rice field. *Indian J. plant physiol.* 33(1):21-26.

Singh, G.R., Pandya, K.S., Choudhary, K.K. and Sharma, R.B. 1997, Biofertilizers in crop production. *Indian Farming.* 46(10):34-36.

Sjamsudin, E., Harjadi, S.S., Poerwanto, P. and Endang, S. 1994. Response of tomato to soil conditioners and NPK dosage on red yellow podzolic soil. *Acta Harticultura.* 369:344-351.

Stewart, W.D.P. 1963. Liberation of extracellular nitrogen by two nitrogen fixing blue-green algae, *Nature.* 200:1020.

Stewart, W.D.P. 1974. Blue-green algae. In: *Biology of Nitrogen Fixation.* (ed.) A. Quispel. North Holland Publishing Co., Amesterdam. pp.202-287.

Stewart, W.D.P., Rowell, P. and Rai, A.N. 1983. Cyanobacteria-eukaryotic plant symbiosis. *Annales de Microbiologie(In stitut Pasteur).* 145B:205-228.

Svircev, Z., Tamas, I., Nenin, P. and Drobac, A. 1997. Co-cultivation of N2-fixing cyanobacteria and some agriculturally important plants in liquid and sand cultures. *Applied soil Ecology.* 6(3):301-308.

Vaishampayan, A. 1998. Physiological responses of genetically improved nitrogen-fixing cyanobacteria to agro chemicalization in relation to paddy culture: Prospects as a source material for engineering herbicide sensitivity and resistance in plants. In: *Advances in plant physiology*. (ed.) A. Hemantaranjan. Scientific publishers, Jodhpur, India. pp. 191-220

Vaishampayan, A., Sinha, R.P., Hader, D.P., Dey, T., Gupta, A.K., Bhan, U. and Rao, A.L. 2001. Cyanobacterial biofertilizers in rice agriculture. *Botanical Review*. 67(4):453-516.

Vazquez, G., Casasola, P.M. and Barrera, O. 1997. Interaction between algae and seed germination in tropical dune slack species: A facilitation process. *Aquatic Botany*. 60:409-416.

Venkataraman, G.S. 1981. All India-Co-ordinated project on algae. Annual report (1980-81). Division of Microbiology, *IARI*, New Delhi, India.

Venkataraman, G.S. 1993. Blue-green algae (cyanobacteria). In: *Biological nitrogen fixation*. (eds.) S.N. Tata, A.M. Wadhwani and M.S. Mehdi. *ICAR*, New Delhi. pp. 45-76

Wagner, G.M. 1997. *Azolla*: A review of its biology and utilization. *The Botanical Review*. 63(1):1-26.

Watanabe, I. and Roger. P.A. 1984. Nitrogen fixation in wetland rice. In: *Current Developments in Biological Nitrogen Fixation*. (ed.) N.S. Subba Roa. Edward Arnold, London. pp. 237-276.

Watanabe, I., De-Datta, S.K. and Roger, P.A. 1987. Nitrogen cycling in wet land rice soil. In: *Advances in nitrogen cycling in agriculture ecosystems*. (ed.) J.R. Wilson. C.A.B. International, United Kingdom. pp.239-256.

Whitton, B.A. 1965. Extracellular products of blue-green algae. *J. Gen. Microbiol*. 40:1-11.

Wolk, C.P. 1973. Physiology and cytological chemistry of blue green algae. *Bacteriol. Rev*. 37:32-101.

4

Biofertilizers : An Essential Component of Organic Farming

PRAVAT KUMAR ROUL AND MOHIMA PRASAD BEHERA

The world population is increasing tremendously. To guarantee enough food for all, use of chemical fertilizers and pesticides become indispensable. The main reason for the increase in per hectare yield in the recent years is the indiscriminate use of chemical fertilizers. But this has created issues such as water pollution, environmental pollution and unsustainable productivity of soil. NPK, the major nutrients required for crop growth are supplied through chemical, fertilizers and organic manures. Chemically fixed nitrogen is a non-renewable source and the largest and costly input in agricultural production. It is very difficult for the small and marginal farmers to purchase and apply recommended dose of inorganic fertilizer at current prices. This crisis has necessitated farm scientist to look for renewable energy source of different inputs in crop production. Bio fertilizers based on renewable energy sources are the cheap and the best cost-effective supplement to enhance production. Sustainable agriculture relies greatly on renewable resources like bio fertilizers. Certain microorganisms are able to fix atmospheric N there by to supplementing nitrogen nutrition whereas certain others help in phosphorus nutrition without application of high levels of chemical fertilizers. The process of biological nitrogen fixation offers an economically attractive and sound means of reducing external nitrogen input. Therefore strategy for improving agricultural production in India should take in to account the use of biofertilizers for N and P nutrition of crop plants.

Biofertilizers are preparation containing sufficient strains of bacteria, fungi or algae capable of fixing atmospheric nitrogen or solubilizing soil phosphorus or stimulating plant growth through the synthesis of growth promoting substances. The current annual production of bio fertilizers in India hardly reaches 22000 tonnes against a potential demand of about 40000 tonnes a year. This indicates the potential of substantial growth in the microbial industry. Biofertilizers are cost effective eco-friendly and renewable source of plant nutrients. Bio fertilizers can be broadly classified into two depending on the nutrients they supply.

BIOFERTILIZERS FOR NITROGEN NUTRITION

Microorganisms such as *Rhizobium, Azotobacter, Azospirillum,* Blue green algae, Azolla etc. supply the nutrient nitrogen through biological nitrogen fixation.

Rhizobium

Legume-Rhizobium symbiosis

The global nitrogen fixation is estimated to be 175 mt per year. Legume -*Rhizobium* symbiosis alone contributes 40 % which accounts for at least 70 mt per year. This symbiosis is in fact a gift of nature to mankind. This meets nearly 60-80 percent of the nitrogen requirement of the host legume. Besides, the cultivation of these will also result in the enrichment of the soil fixed nitrogen. The nitrogen fixation by the legumes inoculated with specific species of *Rhizobium* under ideal and controlled conditions are estimated to be 100-300 in alfalfa, 102-150 in clover 37-196 in cluster bean, 112-152 in peanut, 68-200 in pigeon pea, 49-130 in soybean and 36-63 kg/ha in chickpea. (Subba Rao, 2001). Inoculation with appropriate strains of *Rhizobia* enhance the yield of these legumes (Subba Rao and Tilak, 1977; Sivaprasad and Shetty, 1980; Nair and Sivaprasad, 1982; Sivaprasad and Rai, 1990 and Meenakumari and Nair, 2001). The success of legume - Rhizobium symbiosis however, is dependent on several biotic as well as abiotic factors. Ecological factors like acidity, alkalinity, application of agricultural chemicals and presence of ineffective and antagonistic strains influence the establishment of introduced strains. In order to derive greater benefit from legumes grown in rotations and as inter-crops with nonlegumes, it is necessary that legume seeds are inoculated with specific *Rhizobium.*

TABLE 1

Cross Inocculation Group of *Rhizobium*

Rhizobium species	**Cross inoculation grouping**	**Legume type**
Rhizobium		
R. leguminosarum	Pea group	Pisum, Vicia, Lens
R. phaseoli	Bean group	Phaseolus
R. trifoli	Clover group	Trifolium
R. meliloti	Alfalfa	Melilotus, Medicago, Trigonella
Bradyrhizobium		
R. lupini	Lupini roupg	Lupinous, Orinthopus
R. japonicum	Soybean group	Glysine
R. sp	Cowpea group	Vigna, Arachis
Azorhizobium	Stem nodular	*Sesbania rostrata*

Stem nodulating sesbania biofertilizer

Biological nitrogen fixation by leguminous green manure crops in symbiotic association with *Rhizobium* is a low cost input for rice crop. Many *Sesbania* species are widely used in Asia of which *S. aculeala* and *S. speciosa* are considered to be the most predominant species grown in India (Kannaiyan, 1989). These two are more commonly grown as N_2 fixing green manures because of their vigorous growth habits, higher rates of N_2 fixation and their general ability to withstand a wide range of soil conditions, *viz.*, water logging, salinity and alkalinity (Kannaiyan and Kalidurai, 1995). The stem nodulating tropical legume, *S. rostrata,* has shown better performance as a potential N_2 fixing green manure crop in rice farm ing systems due to -its fast growth rate, N_2 fixing efficiency and flood tolerance (Kalidurai and Kannaiyan, 1989). The use of stem nodulating *S. rostrata* and root nodulating *S. aculeata* as green manure have not only increased the field of rice but also improved the soil microbial population and soil enzymes (Kannaiyan, 1993a).

S. rostrata - Azorhizobium eaulinodans symbiosis is highly species specific. *A. eaulinodans* neither forms effective root nodules in other Sesbania species nor stem and root nodules in any of the stem nodulating *Aesehynomene species* (Ladha et al., 1990). The occurrence of azorhizobia in seeds of *S. rostrata* was reported by Ladha *et al.* (I988) and Balasubramani *et al.* (1992). The transmission of *A. eaulinodans* in stem nodulating *S. rostrata* by a beetle, Mylabris pustulala was reported by Balasubramani and Kannaiyan (1993). Field experiments conducted with fertilizer nitrogen coupled with *S. rostrata* and *S. aculeata* have recorded higher grain yield of rice. Fertilizer nitrogen coated with neem cake along with incorporation of *S. rostrata,* or *S. aculeata* have further increased the grain yield of rice (Kannaiyan, I 993a). Green manures could be fitted into many cropping systems to derive the benefit of the symbiosis for N nutrition. Many leguminous green manures such as Crotalaria, Calapagonium, Clitoria, Desmanthes, Macroptillum etc. have considerable ability to fix atmospheric nitrogen and produce substantial biomass.

Associative Nitrogen fixers

Azospirillum - an associative Nitrogen fixers

Azospirillum spp. are associative nitrogen fixing bacteria which grow in association with various Graminae and other host plants (Subha Rao, 1983; Govindan and Purushothaman, 1985) and cotton and soyabean (Bashanb, 1991). The ability of *Azospirillum* to fix Nitrogen, has been confirmed by many workers. (Oken et al., 1976, Sasikumar, 1996; Kavitha et al. 2003). In addition to nitrogen fixation, *Azospirillum* produces considerable quantity of growth hormones and enhances seedling growth, root development and vigour. Inoculation with *Azospirillum* increased root length, particularly of root elongation gene, increased cell division in the meristematic region of inoculated plants resulted in better root elongation. Inoculated plants showed increase in number of lateral roots with dense root hairs, root volume and root dry weight (Dewan and Subha Rao, 1979). These desirable traits of *Azospirillum* have much application in the production of quality planting materials particularly of spice crops. Pepper cuttings when dipped in the slurry of *Azospirillum* induced better rooting (Yamini Varma, 1995). *Azospirillum* inoculation increased plant height and dry matter production of rice (Karthikeyan, 1981, Sasikumar, 1996), pearl millet (Venkateswarlu and Rao, 1983), pepper (Govindan and Chandy, 1985), Yamini Varma, 1995) chilli (Kavitha 2001). It also induces early seed germination and growth of chilli plants (Kavitha 2003). Growth and vigour of Vanilla, which is the most favoured spice crop nowadays, could be increased by the application of *Azospirillum*. Better sprouting of pepper and Vanila cuttings were noticed, when they were dipped in slurry of *Azospirillum*; The recommended organic manure may also be enriched with *Azospirillum*. In addition to this foliar spraying of *Azospirillum* increases growth and yield of vanilla.

The synergistic effect of *Azospirillum* with other beneficial microorganisms has been well established. The beneficial effect of mixed inoculum of *Azospirillum* and *Azotobacter* was observed in rice, maize, sorghum, wheat etc. Similarly, seed inoculation with *Azospirillum* and *Rhizobium* increased the dry matter and nitrogen content of chick pea, cowpea etc. Combined inoculation of phosphate solubilizing bacteria and *Azospirillum* increased dry matter and lipid in sorghum, bajra and cotton. Application of AMF together with *Azospirillum* has been found to benefit crop growth and yield (Subbiah, 1994; Rajput, 1999).

***Azospirillum* biofertilizer for rice**

Field experiments conducted at different places in India have indicated the positive response of rice cultivars to *Azospirillum* inoculation. (Subba Rao, 1979; Roa *et al.,* 1983). Experiments conducted at IRRI, Philippines showed that inoculation with *Azospirillum* lipoferum promoted early tillering and also the growth of rice and sgnificantly increased filling rate of grain and the grain weight per plant at harvest (Watanabe and Lin, 1984; Nayak *et al..,* 1986). Response of rice to inoculation of *A. lipoferum* has been reported by several researchers (Rao *et al.,* 1983; Kumar and Balasubramanian, 1986; Gopalaswamy and Vidhyasekaran, I989; Jeyaraman and Purushothaman, 1988). However, the response of rice to *Azospirillum* inoculation under low land rice field condition was variable, depending on the initial nitrogen status of the soil and the varieties of the rice. Beneficial effect of *Azospirillum* inoculation have been reported on maize, sorghum, sunflower, and pearl millet (Dhanabal *et al.,* 1978); sorghum (Subba Rao *et al.,* 1980, 1986; Okon and Lababdera-Gonzalez, 1994); pearl millet (Subba Rao et aI., 1986; Tilak and Subba Rao, 1987); maize (Tilak *et al.,* 1982; Fulchieri and Frioni, 1994) and cotton (Purushothaman and Gunasekaran, 1981).

Free living nitrogen fixers

Azotobacter - a free living Nitrogen fixer

Azotobacter are free living nitrogen fixing bacteria, which are also known to produce growth promoting substances which stimulate germination of seeds, root development and accelerate plant growth. Significant increase in yield due to *Azotobacter* inoculation has been obtained in crops like rice, wheat, brinjal, tomato etc. (Balasundaram and Abhiswar Sen, 1971; Mehrotra and Lehri, 1971; Joy and Shinde, 1976). Enhanced seed germination observed may be due to the ability of *Azotobacter* to produce growth substances and anti-fungal antibiotics (Leksmikumari et al. 1975). The problem with *Azotobacter* is that substantial amount of energy is required by the bacterium for nitrogen fixation. In soils, poor in organic matter, *Azotobacter* runs short of energy and ceases to fix nitrogen. This particular non-symbiotic bacterium fixes 20-25 kg N per ha. High amount of organic mulch and adequate level of moisture are to be maintained for proper growth and N fixation. In chillies, Azotobacter inoculation with half nitrogen level could give significant growth improvement and yield.

Azolla biofertilizer for rice

Azolla is a free-floating waterfern which in symbiotic association with the cyanobacterium *Anabaena azollae*-contributes substantial amount of biologically fixed nitrogen to rice crop. *Azolla* has been used as a biofertilizer for rice crop in India, China, Vietnam, Thailand, Philippines, Korea, Sri Lanka, Bangladesh, Pakistan, Nepal, Burma, Indonesia, Brazil and West Africa, but the potential of this biological system has not been fully exploited. The *Azolla - Anabaena* association is indeed a "living floating nitrogen manufacturing factory" which utilizes the energy from photosynthesis to fix atmospheric nitrogen amounting to 150 to 200 kg ha^{-1}, *Azolla* is used as biofertilizer for rice crop and it could contribute 40-60 kg Nha per crop (Kannaiyan, 1992). The important factor in using *Azolla* as a biofertilizer for rice crop is its rapid decomposition in soil and quick release of its nitrogen to rice plants (Kannaiyan, 1990).

An increase in the grain yield of rice by the application of *Azolla* as green manure, dual crop or both with rice was equivalent to 30-40 kg N ha^{-1} and this has been well documented by several investigators in, various countries such as China (Liu Chung Chu, 1987), Philippines (Mabbayad, 1987; Watanabe, 1987), India (Singh, 1977; Kannaiyan,

1987), Thailand (Loudhapasitiporn and Kanareugsa, 1987), Sri Lanka (Kulasooriya et al ., 1987), USA (Talley et al., 1977) and West Africa (Diara et al., 1987). The use of *Azolla* in rice field considerably increased the soil microbial as well as enzymes activity (Kumar et al., 1995). It is an important low cost input, which plays a vital role in improving soil quality in sustainable rice farming (Kannaiyan, 1992; 1993a and 1995). Gopalaswamy and Kannaiyan (1998) have investigated the potential of *Azolla* hybrids and reported that inoculation of *Azolla* hybrids significantly improved the soil organic carbon besides increasing the rice yield.

Nitrogen fixing blue green algae (BGA)

The species of blue green algae which are known to fix atmospheric nitrogen are classified into three groups (i) heterocystous - aerobic forms, (ii) aerobic unicellular forms and (iii) nonheterocystous, filamentous, microaerophilic forms. The composite inoculum of BGA consisting of *Nostoc, Anabaena, Calothrix, Tolypothrix, Plectonema, Aphanothcae, Gloeocapsa, Oscillatoria, Cylindrospermum, Aulosira* and *Scytonema* have been used for inoculation in rice (Kannaiyan, 1985). A technology for mass scale production of composite culture of BGA under rice field condition was developed by Kannaiyan(1985). These N_2 fixing cultures have better adaptability and profitability in wetland rice. Fertility of tropical rice fields is attributed to the inherent ability of cyanobacteria to fix nitrogen and sustain fertility of rice fields.Besides, improving soil fertility, increasing soil organic matter and production of growth substances, an increase in gram yield up to 10-20% was observed as result of algal application (Venkataraman, 1981).

Inoculation of BGA is highly suitable during dry season. Application of dried soil based BGA flakes at the rate of 10 kg per hectare is recommended for rice crop. Five hundred gram of single super phosphate may be mixed With 10 kg of soil based BGA and applied 10 days after transplanting in rice field. Continuous application of blue green algae to the rice fields · for at least 3 or 4 consecutive seasons sustain the crop yield at reduced levels of commercial nitrogen fertilizers (Kannaiyan, 1993 b). The beneficial effect of algal inoculation on the grain yield of rice has been well documented at a number of locations (Venkataraman, 1972 and Kannaiyan, 1985). Algal inoculation to rice crop was effective in different agro-climatic condition and soil types (Goyal and Venkataraman, 1971). The soil based BGA inoculum could survive for more than 2 years (Kannaiyan, 1990).

BIOFERTILIZERS FOR PHOSPHOROUS NUTRITION

Phosphorous solubilisers

Next to nitrogen, phosphorus is the important nutrient, which is required for plants. Most of the soil phosphorus is in unavailable form and hardly about 1-2% of it is incorporated into the above ground parts of the plants. Generally, phosphorus is supplied to the soil, in the form of chemical fertilizers or organic source such as decomposed plant and animal materials . A few micro organisms are capable of solubilizing insoluble form of P in the soil and play important role in Phosphorus nutrition of crop plants. Introduction of efficient P solubilizers brings about dissociation of bound forms of P and increase availability to the plants. Several soil bacteria such as *Bacillus, Pseudomonas* and fungi like *Aspergillus* and *Penicillium* secrete organic acids and chelating agents and bring about solubilization of bound phosphates (Sundar Rao and Sinha, 1963). Beneficial effect due to inoculation with phospho-bacteria was reported for different crops under diverse agro-climatic conditions from India (Gaur, 1990). A saving of 50 % of the recommended level of P_2O_5 is possible with the inoculation of phosphobacterium (Kathiresan et al. 1995). The beneficial effects of phosphate solubilising microbes in increasing crop yields as well as in reducing the requirement of phosphate fertilizers has been conclusively proved (Gaur, 1990). The synergistic effect of combined inoculation of nitrogen fixing *Rhizobium* and PSM was obtained in soyabean (Natarajan and Gunasekharan, 1991), peas (Natarajan, 1980) and groundnut (Natarajan and Subramanian, 1995). Combined inoculation of PSM and azospirillum on crops like sorghum, (Alaguwadi and, Gaur 1992), bajra (Nirmala and Sundaram, 1996) and cotton gave significant increase in dry matter and yield. For the organic production of turmeric application of biofertilizers like phosphobacteria @ 7.5 kg/ha is recommended (Ravikumar,2002). However, it is yet to become an accepted agron organic practice among the farmers of Kerala and the efforts are required in this direction.

Arbuscular Mycorrhizal Fungi (AMF)

Mycorrhizal association is the mutualistic symbiosis, occurring between certain soil fungi and roots of higher plants. Nearly 83% of dicotyledons and 13% of monocotyledons form mutualistic symbiosis with Arbuscular Mycorrhizal Fungi of the, endo mycorrhizal group

(Trappe, 1987). They are obligate symbionts and are not host specific. These are formed by a group of zygomycetous fungi belonging to the order Glomales. (Rosendahl et al. 1994). The beneficial effects of AMF association in the host nutrition in the uptake of P are well documented (Mjosse, 1973; 1997; Abdul Khaliq et al.2001).

Mycorrhizal infection enhances plant growth by increasing nutrient uptake either by increasing the absorbing surface and mobilizing sparingly available nutrient sources or by secretions of ectoenzymes (Bolan, et al. 1987). AM plants absorb and accumulate more P than non AM plants either due to more efficient translocation coupled with better exploration of soluble P by the fungal hyphae or due to solubilization of insoluble P by the AMF. The role of AMF in the uptake of other nutrients Viz. Cu, Zn, Ca, K, Fe, Mn is also well documented. (Abdul - Khaliq et al. 2001). Further AM association increases the uptake of water (Safir et al. 1972), reduces stress responses of plant to toxicity and drought (Atkinson and Davidson, 1972, Guttay, 1976) improve soil texture and reduce soil erosion (Clough and Sutton, 1978) and enhances revegetation of degraded soils like mine soils (Pfleger *et al.* 1994). The enhanced uptake of P and other nutrients has been well documented in crops like cassava, cardomom, ginger, turmeric, cashew and legumes grown in P fixing soils of Kerala (Sivaprasad et al, 1984,1990,1992,1998: Geethakumari et al 1994). The most remarkable consequence of AMF colonization is the enhanced growth and yield. Extensive studies have been conducted at different locations which emphasise the importance of AMF technology. AMF inoculation increased growth and yield of pepper, ginger and turmeric (Sivaprasad, 1998).

Biofertilizers are harmless, improve soil physico-chemical and biological properties. Secrete growth promoting substances, hormones and vitamins. They help in better nutrient uptake and increased tolerance towards biotic stress. The waste lands can be enriched and productivity may be increased by application of biofertilizers. The economic benefit to cost of biofertilizer is always higher so that their use is well accepted by the farming communities.

REFERENCES

Abdul-Khalique, Gupta, M.L. and Kumar, S. 2001. The effect of VAM fungi on growth of pepper mint. *Indian Phytopath.* 54 : 82-84.

Balasubramani, G. and Kannaiyan, S. 1993. Transmission of *Azorhizobium caulinodans* in stem nodulating *Sesbania rostrata* by Mylabris pustulata. J. Soil Biol. Ecol., 13: 75"77.

Balasubramani, G., Krishnaveni, S. and Kannaiyan, S. 1992. Inhibition of *Azorhizobium* and *Rhizobium* by seed leachates. World .r. Microbiol. Biotechnol., 8 : 542-543.

Dhanabal, N., Purushothaman, D. and Nadanam, M. 1978. Effect of seed inoculation with *Spirillum lipoferum* on pearl millet and sorghum. Food Fong. Agrl., 10: 85-88.

Diara, H.P., Van Brandt, H., Diop, A.M. and Van Hove, C. 1987. Azalia and its use in rice culture in West Africa. In : Azalia Utilization. Proc. Workshop on Azalia use. Fuzhou, Fujian. China, IRRI, Los Banos. Philippines. pp. 147-152.

Fulchieri, M. and Frioni, L. 1994. Azospirillum inoculation on maize (*Zea mays*): Effect of yield in a field experiment in Central Argentina. Soil Biol. Biochem., 26: 921-924.

Gaur, A.C. 1990. Phosphate solubilising microorganisms as biofertilizers, Omega Scientific Publishers, New Delhi, p. 176.

Gopalaswamy, G. and Vidhyasekaran, P. 1989. Response of different rice cultivars to *Azospirillum* inoculation. Cur. Sci., 58 : 752-754:

Gopalaswamy, G. and Kannaiyan, S. 1998. Effect of inoculation of *Azolla* hybrids on soil organic matter and rice yield. In: S.Kannaiyan (Ed) Cyanobacterial biofertilizer for rice crop. Tamil Nadu Agricultural University, Coimbatore, India. pp. 214-217.

Goyal, S.K. and Venkataraman, G.S. 1971. Response of high yielding rice varieties to algalization. II. Interaction of soil types with algal inoculation. Phykos, 10: 32-33.

Jeyaraman, S. and Purushothaman, S. 1988. Biofertilizers efficiency in lowland rice. IRRN., 13: 24-25.

Joseph, P.J. 1997. Management of rhizome rot of ginger using VA Mycorrhizae and antagonists. Ph.D. thesis, Kerala Agricultural University, Thrissur, 192 pp.

Kalidurai, M. and Kannaiyan, S. 1989. Effect of *Sesbania rostrata* on nitrogen uptake and yield of lowland rice. J. Agron. Crop Sci, 163: 284289.

Kannaiyan, S. 1985. Studies on the algal application for lowland rice crop. Tech. Bull, Tamil Nadu Agricultural University, Coimbatore, India. p. 27.

Kannaiyan, S. 1987. Azolla Biotechnology. Tech. Bull, Tamil Nadu Agricultural University, Coimbatore, India. p10.

Kannaiyan, S. 1989. Nitrogen fixing potential of stem nodulating *Sesbania rostrata*. In: Organic recycling in Asia and Pacific. PAPA Bull., 5.

Kannaiyan, S. 1992. Azolla biofertilizer technology for rice. Tech. Bull., Tamil Nadu Agricultural University, Coimbatore, India. p. 56.

Kannaiyan, S. 1995. Biofertilizers for rice: Status and scope. In: S. Kannaiyan (Ed) Rice Management Biotechnology, Assoc. Pub. Co., New Delhi. pp. 237-265.

Kannaiyan, S. and Kalidurai, M. 1995. Nitrogen fixing potential of stem nodulating *Sesbania rostrata* and its effect on rice yield. In: S.Kannaiyan (*Ed*) Rice Management Biotechnology. Assoc. Pub. Co., New Delhi. pp. 173-195.

Kannaiyan, S. 1993a. Biofertilizers for rice. Tamil Nadu Agricultural University, Coimbatore, India. p. 60.

Kannaiyan, S. 1993b. Algal biofertilizer technology for rice. Indian Phyco. Rev., 2 : 73-82.

Kannaiyan, S. 1990. Biotechnology of biofertilizer for rice crop. Tamil Nadu Agricultural University, Coimbatore, India. p. 212.

Kulasooriya, S.A., Hirmburegama, W.K. and W.Abeysekara, S. 1987. Use of Azolla in Sri Lanka. In : Azolla Utilization. Proc. Workshop on Azalia use. Fuzhou, Fujian, China, IRR!. Los Banos, Phillipines. pp. 131-140.

Kumar, K. and Balasubramanian, A. 1986. Field response of rice to Azaspiril/um biofertilizer Curro Res., 15: 74 -76.

Kumar, K., Thangaraju, M. and Kannaiyan, S. 1995. Use of nitrification inhibitors and nitrogen use efficiency in lowland rice soil. In: S. Kannaiyan (Ed) Rice Management Biotechnology, Assoc. Pub. Co., New Delhi. pp. 101-110.

Ladha, J.K., Pareek, R.P. and Becker, M. 1990. Stem nodule symbiosis and its unusual properties. In : P.M.Gresshoff, L.E. Roth, G.Stacey and W.L. Newton, (Eds) N_2 fixation: Achievements and Objectives. Chapman and Hall, New York, London. pp. 633-640.

Ladha, K.R., Kannaiyan, S. and Subramanian, S. 1988. Uptake of nitrogen in different growth phases of rice through *Azalia* biofertilizer. Acta Agron. Hungarica, 37: 239-243.

Lekshmikumari, M., Kavimandan, S.K. and Subba Rao, N.S. 1976. Occurance of nitrogen fixing spirillum in roots of rice, sorghum, maize and other plants. *Indian J. Exp. Biol.* 14 : 638-639.

Liu Chung Chu. 1987. Re-evaluation of Azolla utilization in agricultural production. In: Azalia Utilization. Proc. of the Workshop on Azolla use. Fuzhou, Fujian, China, IRRI, Los Banos, Philippines. pp.67-76.

Loudhapasitiporn, L. and Kanareugsa, C. 1987. Azolla use in Thailand.

In : Azolla Utilization Proc. on the Workshop on Azolla use. Fuzhou, Fujian, China, IRRI. Los Banos, Philippines. pp.119-122.

Mabbayad, B.B. 1987. The Azalia program of the Philippines. In :/zol/a Utilization. Proc. of the Workshop on *Azolla* LIse. Fu71lOU. Fujian. China. IRRI, Los l3anos. Philippines. Pp. 101- 108.

Meenakumari, K.S. and Nair, S.K. 2001. Influence of soil chemical properties on root nodulation by *Bradyrhizobium* sp. In cowpea, blackgram and greengram. *Journal of Tropical Agriculture* 39 : 62-64.

Nayak, D., Ladha, N.J.K. and Watanabe, I. 1986. The fate of marker *Azospirillum lipoferum* inoculated into rice and its effect on growth, yield and N2 feeding and 15N dilution techniques. Biol. Fert. Soils, 2: 7-14.

Okon,Y. and Labandera-Gonzalez, C.A. 1994. Agronomic applications of *Azospirillum;* An evaluation of 20 years worldwide field inoculation. Soil Biochem, 12: 1591-1601.

Purushothaman, D. and Gunasekaran, S. 1981. *Azospirillum,* A biofertilizer forcotton. TNAU Newsletter, 11:4.

Rao, V.R., Nayak, D.N., Charyulu, P.B.D. and Adhya, T.K. 1983. Yield response of rice to root inoculation with *Azospirillum*. J. Agric. Sci., 100: 689-691.

Ravikumar, P. 2002. Production technology of organic turmeric. *Spice India* Feb. 2-4, Vol No. 15, No. 2

Sasikumar S. 1996. Development of inoculate cultures of *Azospirillum* for rice (*Oryza sativa* L.) in Kuttanadu M.Sc. (Ag) thesis, Kerala Agricultural Unviersity, Thrissur.

Singh, P.K. 1977. The use of *Azolla pinnala* as green mariure for rice. IRRN, 2 : 7.

Sivaprasad, P. and Rai, P.V. 1990. Synergistic effect of *Glomus fasciulatum* and Rhizobium spp. On *Cajanus cajan* (L.) Mills P. under glass house and field conditions. *Indian J. Agric. Sci.* 61 (2) : 468-469.

Subba Rao, N.S. 1979. Biofertilizer, in Indian Agriculture: Prablems and Prospects. Fert. News. 24: 84-90.

Subba Rao, N.S. and Tilak, K.V.B.R. 1977. Rhizobial cultures and their role in pulse production souvenir Bulletin, Directorate of Pulses Development, Lucknow. India pp. 31-34.

Subba Rao, N.S., Tilak, K.V.B.R., Singh, C.S. and Nair, P.V. 1986. Yield and nitragen gains af sorghum as influenced by *Azospirillum brasilense.* In: S.P.Wani (Ed) Cereal nitragen fixatian. ICRISA T, Patancheru, India. pp. 69-74.

Subba Rao, N.S., Tilak, K.V.B.R., Singh, C.S. and Nair, P.V. 1986. Yield and nitrogen gains af sorghum as influenced by *Azospirillum brasilense.* In: S.P.Wani (Ed) Cereal nitrogen fixation. ICRISA T, Patancheru, India. pp. 69-74.

SubbaRaa. N.S., Tilak, K.V.B.R. and Singh, S. 1980. Root nodulation studies an *Aeschynomene aspera.* Plant Sail, 56: 491·494.

Sundara Rao, W.V.B. and Sinha, M.K. 1963. Phosphate dissolving organism in soil and rhizosphere. *Indian J. Agric. Sci.* 33 : 272-278.

Talley, S.N., Talley, B.J. and Rains, D.W. 1977. Alexander Hollaender (Ed) Genetic engineering for nitrogen fixation. Plenum Publishing Company, New York, USA. pp. 250-282.

Tilak, K.Y.B.R, Singh, C.S. and Subba Rao, N.S. 1982. *Azospirillum* and *Azotobacter* inoculation: Effect on yield of maize and sorghum. Soil Biol. Biochem., 14: 417-418.

Tilak, K.V.B.R. and Subba Rao, N.S. 1987. Association of *Azospirillum brasilense* with pearl millet (Pennisetum americanum). Biol. Fert. Soils, 4: 97-102.

Venkataraman, G.S. 1972. Algal biofertilizers and rice cultivation. To-day and Tomorrows Printers, New Delhi. p. 75.

Watanabe, I. 1987. Fertility and fertilizer evaluation for rice. In : *Azolla* Utilization. Proc. of the Workshop on Azolla use. Buzhou, Fujian, China, IRRI, Los Bailos, Philippines. pp.197-208.

Watanabe, I. and Lin, C. 1984. Response of wetland rice to inoculation with *Azospirillum lipoftrum* and *Pseudomonas* sp. Soil Sc. Plant Nutr., 30: 117-124.

❑❑❑

5

Organic Farming for Sustainable Agriculture Development

DHIMAN MUKHERJEE

"An organic farm, properly speaking, is not one that uses certain methods and substances and avoids others; it is a farm whose structure is formed in imitation of the structure of a natural system that has the integrity, the independence and the benign dependence of an organism" Joy "The Gift of Good Land"

INTRODUCTION

In the process of attaining higher levels of food production for matching the demand of growing population during the past four decades, emphasis was laid on intensive agricultural practices. With the increase in crop yields from modern farming techniques reaching a plateau in most of the countries and the environmental problems due to excessive use of chemical fertilizers and pesticides becoming a matter of concern (Mukherjee, 2003), the need for sustainable agriculture is increasingly being felt, the world over. In the view of resurgence of interest in alternate in a recent year, organic farming has been considered to be a sound and viable option in most of the countries. In this direction, the recommendations of the Atlanta Conference (1981) on "Organic Farming" have acted as catalysts in trigerring interest in the organic agricultural systems across the world (Dahama, 1997). Further, in order to cope with the problem of pesticide and herbicide toxicity in various food commodities, the Food and Agriculture Organization (FAO) proposed "The World Food Summit

Plan of Action (1999)" in recognition of the importance of developing alternative sustainable agriculture practices such as organic farming. The goal of the Action Plan was to reduce environmental degradation while creating income from the farming operation. Organic farming is an integrated farming system which involves both technical aspects (soil, agronomy, weed, and pest management) and economic aspects (input, output, and marketing) as well as human health. They seek to raise food in accordance with basic biological truths (Long, 2002). The goals of organic farming include maintaining a biodiverse crop, keeping soil healthy and fertile for future years; minimizing pollution, producing high quality products; avoiding synthetic fertilizers and pesticides in agriculture, recycling nutrients through biological interactions and encouraging ecological wholeness (Halweil, 2001). The sustainable agricultural system that survived for thousands of years in the east were those, which were able to maintain the regular cycle of soil organic matter (SOM) in their soil. This is done with the integration of the farming system with livestock, forestry and green manure. Once this cycle was broken by adoption of industrial agriculture, it led to depletion of the SOM resulting in poor quality of food, diseased soil and adding emission (Hegde, 2009).

REASON FOR ORGANIC FARMING

Intensification of agriculture through massive adoption of high yielding cultivation, increased use of synthetic inputs like chemical fertilizers and pesticides, greater exploitation of irrigation potentiality of surface and ground water resources and farm mechanization have largely been responsible for a spectacular achievement in the food grain production that we have achieved over past three decades. Paradoxically however overexploitation of natural and renewable resources and indiscriminate and irrational use of synthetic inputs like inorganic fertilizers and pesticides in view of producing more and more from unit piece of land are being increasingly realized to seriously impair the ecological balance and putting the environment in jeopardy. High yielding cultivation is more fertilizer responsive which often led to aggravation of pest problem as the plants become succulent enough to be fed upon by a variety of crop pests. This in turn necessitates increasingly huge amount of pesticides to combat pest problem. Increased use of pesticides has emerged as a potential source of danger to sustainability of environment that endangers the existence of all forms of life on this planet. Perils and pitfalls of pesticides have been well evidenced due to their residual toxicity in our food chain. From a number of trials conducted across the country, toxic residues of pesticides have been revealed in the food stuff not only of plant but also of animal origin like milk and milk products, fish, meat and egg etc. at concentrations much higher than the permissible level of human body (Lipson, 2001). Therefore the apparent contradiction of our necessity for nutritional security on one hand and environmental sustainability on the other makes it inevitable to resort to the organic or eco-farming system as it appears to be a possible option to meet both these objectives. The later implies a farming system that primarily aims at cultivating land and raising crops under ecologically favorable condition. Addition of organic manures such as FYM, recycling of organic wastes through composting, green manures and biological inputs like vermicomposts and biofertilizers etc. constitute important components for plant nutrient management in organic farming. Similarly it also takes utmost advantage of the natural mechanism for pest management with utilization of bioagents such as predators and parasites available in nature in plenty and the botanical pesticides which are effective in controlling crop pests posing no risk to the environment (Halweil, 2001). Agronomic practices such as crop rotation with judicious selection of crops, inter cropping and companion cropping, stubble mulching and use of resistant varieties are among the important factors contributing to organic farming (Mukherjee, 2005).

SUSTAINABLE AGRICULTURE AND ORGANIC FARMING

Sustainable agriculture encompasses many different production methods, systems, and approaches that aim to meet the goals of profitability, stewardship, and quality of life (Bourn and Prescott, 2002). One of those approaches, but by no means the only one, is organic farming. Organic agriculture is generally considered to be under the sustainable agriculture "umbrella." But it is not exactly a subset, since organic practices may conflict with sustainability goals in certain situations. However, keeping present situation in mind, organic farming is a key for sustainable agriculture.

Organic farming is a production system for growing crops that avoids the use of most man-made fertilizers and pesticides, applications of sewage sludge, genetic engineering, and radiation. In their place, animal manure, crop rotation, traditional plant breeding, and other materials and strategies are used to maintain economic crop yields, manage pests, and preserve environmental quality. Ultimately these objectives fulfill the sustainability objective (Singh *et al.*, 2004). Organic farming has been one of the fastest growing sectors within production agriculture in the India and several other industrialized countries, and there is little indication that the demand for food grown organically will decline anytime soon (Hunter, 2002). This kind of farming uses production strategies that consider a holistic, systems-based management approach results in food that is healthier to eat and an environment that is safer to live in. There are assumptions throughout the organic literature of differences between organic and conventional systems with respect to their effects on soil physical properties, soil insect fauna and nutrient flow within the soil, crop health and nutritional value of the harvested crop.

NEED OF ORGANIC FARMING FOR SUSTAINABLE DEVELOPMENT

Organic farming need to the following problem, which are in present context very important as these are becoming threat to our future generation.

1. Adverse effect of agro-chemicals on soil, water, food and atmospheric environment

Adverse use of agro chemicals takes place in modern farming by many developed and under develpoed countries, there are various adverse effect on soil, water, food and atmospheric environment (Mukherjee, 2006). The increased use of various pesticide in recent years causes more concern about their effect on farmer's health. While herbicide technology has made remarkable progress in terms of developing safe herbicides that are less toxic to human beings, may farmers still suffer from chemical poisoning after applying herbicides. The extensive use of pesticides posses a more direct problem on farm workers. Epistemologies data shows that workers who handle pesticides more than 20 days a year have an increased risk of developing certain type of cancer (Dahama, 1997).

2. Fertilizer pollution

In the developed countries, there has been intensive fertilizer use for the last four decades. If the polluting effects of fertilizers are being observed now, similar problems in developing countries should be expected in the near future. Some important problems associated in fertilizers pollution are summarized below.

(1) Nitrate Pollution : Application of N_2 fertilizers such as urea and ammonium sulphate to soils produces acid by two processes. Firstly, the natural process of oxidation of ammonia ions to nitrate ions release acid. Part of acid produced is neutralized by alkaline ions released by plants during the subsequent uptake of the nitrate ions. Secondly, since nitrate ions are not strongly absorbed by the soil they are liable to leach or move down through the soil. The negatively charged nitrate ions carry positively charged basic cations such as Ca, K, Mg and Na in order to maintain the electric charge on the soil particles. A high nitrate concentration indicates likely presence of harmful bacteria as well. In condition, to high enrichment, NO_3 may produce as state known as methamoglobinema (blue babies) which generally affects the infants under six months of age. Repeated heavy dose of nitrate on ingestion may likely to cause carcinogenic diseases. Consumption of high dose of NO_3 may develop symptoms of dizziness, abdominal cramps, vomitting, weakness, convulsion, mental impairment and even nitrosamine which cause stomach cancer. Apart from this, over use of N_2 fertilizers leads to swindling of earthworms from the particular area, earthworms have always been considered a farmers friends and their absence mean loss to the soil fertility.

(2) Accumulation of Heavy Metals : Contamination of soil by heavy metal through fertilizers such as cadmium from phosphatic fertilizers is also getting increasing attention of environmentalists (Mukherjee, 2004). Health hazards associated with heavy metals entering the food chain through soil is demanding attention. Fertilizers contain heavy metals as impurities. The application of rock phosphate or its produce to soil always implies the addition of significant amount of lead and cadmium into the soil. Analysis of several commercial fertilizers commonly used revealed that a combination of low analysis and straight fertilzers can add more lead and cadmium to soil than high analysis and mixed fertilizers (Singh and Sekhon, 2002).

3. Eutrophication of water

It is the process of enrichment of surface water bodies like lakes, reservoirs and dreams with nutrients. Nutrient enrichment of water bodies results in intense proliferation and accumulation of algae and higher aquatic plants in excessive quantities which can result in detrimental changes in water quality and can significantly interfere with the use of water resources. The excessive accumulation of dissolve nutrients such as phosphorus, nitrogen, silicon and other element in water leads to an excess production of algal biomass. This requires a corresponding increased supply of oxygen for decomposition of organic materials when the algae and their remnants sink to bottom, reduced O_2 content and eventually anaerobic condition may prevail. This is syptomatic a severe eutrophication and may leads to a serious loss of marine life, blockage and aquatic passages and a major reduction in real estate value of the affected areas. The normal N:P ratio in water is around 20:1. Algal growth increase when this ratio drops to around 7:1 (Asmed, 1993). In the developed countries like USA due to heavy use of fertilizers, more than 50% of the community and rural wells are contaminated with nitrate nitrogen (NO_3N). The contamination has exceeded the standard (10mgNO_3/lit.) of drinking water by 2% (Dahama, 1997).

4. Stratospheric changes

Many research studies involved in the atmosphere may increase the rate of reaction in the stratosphere that to be the distraction of ozone layer. The stratophere ozone layer shields the biosphere from the harmful ultraviolet radiation and also influences the earth's temperature (Dahama, 1997). Nitrogen loss to the atmosphere through denitrification may contribute to **"green house gases"** in the atmosphere thereby exacerbating the problems of the breaking down of ozone layer. Nitrogen losses can be particularly high from intensively cultivated and fertilized land whether the fertlilzer is from organic or inorganic (Mukherjee, 2007).

5. Pesticide pollution

By the use of different types of poisonous subtances as pesticide, it contributes towards imbalances in the ecosystem and polluting the environment. Pesticides are now extremely used to control various pests which are harmful to the crops raised by man for food, feed and fibre production. However, most of the chlorinated pesticides are non-biodegradable and leave residue which are detrimental human beings, animals and the environment.

(1) Insecticide Pollution : The presence of residues of insecticide in food commodities and other component of the environment is a matter of serious concern. Even small quantities of the residues ingested daily along with food can build up to high levels in body fat (Dhaliwal and Arora, 1998). There is therefore a contant fear that society is being slowly poisoned by the intake of food contaminated with pesticide residues. During the 1900's, there was little information available on the band imposes of pesticide on soil and water quality. Organochlorine insecticides such as chlorine, dieldrin, DDT and heptachlor were widely used in agriculture to control insect/pests in different countries like Nepal. Since 1960's, the organochlorine use was progressively restricted and finally banded now. Nevertheless, their residues still found in soil and continue to cause problems of food and food contamination. The metabolite of DDT still occur in the top 150 mm of the soil profile Centerbury region in Switzerland (Bould, 1994).

(2) Herbicide Pollution : The use of chemicals for controlling weeds started with the introduction of 2,4-D in 1940's. The usage of herbicide is higher than any agro-chemicals (Mukherjee, 2003). Farming has now realized the importance of herbicide usage for harvesting higher crop yields. But presently herbicide plays an important role for the disturbance of soil eosystem when soil micro-flora and fauna lies in the breaking down of organic matter, incorporating it into the soil and releasing nutrients for plant growth. The herbicide can have direct effect upon decomposing micro-organisms, rhizosphere micro-organisms, root pathogens and disease antagonists such as parasites and predators as well as organisms pathogenic to invertebrates. Herbicides are designed specifically to minimize plant diversity by controlling weeds this promoting monoculture. So, they can also indirectly decrease populations and diversity of related soil organisms and lessens the natural input of organic matter into soil as well as have direct effect on soil organism. Doses of herbicide caused to phytotoxic effect not only on human health but also to crop cultivar too (Mukherjee and Singh, 2006). So one can approach organic way of cultivation for various weed control methods.

PRINCIPLES OF ORGANIC FARMING

The major widely used term **"organic farming"** describes two major aspects of alternative agriculture.

A. The substitution of manure and other oranic matter as organic fertilizers.

B. The use of biological pest control instead of chemical pest control.

Organic farmers emphasize using only organic fertlilzers for fertility maintenance. In many aspects, inorganic farming is a way of life as it is a method of farming. The profitability of organic farms depends on the higher prices that their products demand in the market place. Organic agriculture is not based exclusively on short term economics, but also considers ecological concepts. It utilizes appropriate technology and appropriate traditional farming methods. This form of farming can also be called sustainable form of farming or sustainable agriculture. The principles of this method are:

- Organize the production of crops and livestock and the management of farm resources so that they harmonize rather than conflict with natural sysem.
- Use and develop appropriate technologies based upon an understanding of biological systems.
- Achieve and maintain soil fertility for optimum production by relying primarily on renewable resources.
- Use diversification to pursue optimum production.
- Aim for optimum nutritional value of staple food.
- Use decentralized structures for processing, distributing and marketing of products.
- Strive for equitable relationship between those who work and live on the land.
- Maintain and preserve wildlife and their habitats.

COMPONENTS OF ORGANIC FARMING FOR SUSTAINABLE AGRICULTURE

1. Crop and soil management

Organic farming system encourages the use of rotations and measures to maintain soil fertility. Carefully managed soil with a high production of humus offer essential advantages with respect to water retention ion exchange, soil erosion and animal life in the soil. Green manuring and inter-cropping of legumes is another important aspect for biological farming systems not only in regard to weed control but also in reducing the leaching of nutrients and in reducing soil erosion. A green cover throuthout most of the year is one of the main goals of such farming methods. Depending on the green manure mixture or the legumes used for under sowing, there may be an increased soil organic matter and soil N_2 as well as in other nutrients.

2. On-farm waste recycling

Increase price of chemical fertilizers have enables organic wastes to regain an important role in the fertilizer practices on the farm. Good manure management means improved fertilizers value of manure and slurry and less nutrient losses. Composting of all organic wastes in general and of Farm Yard Manure (FYM) or feedlot manure in particular is important in organic farming.

3 Non-chemical weed management

Weed management is one of the main concerns in organic agriculture. Generally, all aspects of arable crop production play an important role in a system approach to problems. The elements to consider in preventing weed problems are crop rotation, green manuring, manure management and tillage (Singh, 2004). Mulching on a large scale by using manure spreaders may also be useful in weed control. Crop genotype differs in their ability to suppress weeds (Fisher *et al.*,1997). More competitive crop varieties and higher seed rate can smoother weeds and this is an effective method of weed control in cropping field.

4. Domestic and industrial waste recycling

Sewage and sludge use for crop production can form an important component of organic farming if treatment and application methods are improved further.

5. Energy use

In the energy requirement for production measured per rupees of produce for organic farms is only one third of what it is for their conventional counterparts. Because N-fertilizer and pesticides are not used by biological farmers, the comparison of total energy input/ha. with total energy output favors biological farming systems.

6. Food qualities

Food quality is one of the main issues, which concerns both scientists and consumers. Nitrates in water and farm produce, desirable components, pesticides residues, keeping quality and physiological imbalances are some of the important aspects of food quality. Biologically produce vegetable and fruit contain lower concentration of nitrate and pesticide residues as compared to those produced using chemical input (Mukherjee, 2002). Further, it has been shown that not only a decrease in the concentration of undesirable component but also an increase desirable component (such as vitamin C, iron etc.) can be found when produce is grown organically (Singh and Sekhon, 2002).

7. Ecological agriculture

The growing concern about environmental degradation, dwindling natural resources and urgency to meet the food needs of the increasing population are compelling farm scientist and policy makers to seriously examine alternative to chemical agriculture.

8. Biodynamic farming

In the organic production system, in general micronutrient are not taken care of, there is every doubt that over long duration, their deficiency may create production constraints and these technology might be a failure rather than a sustainable alternative. Biodynamic agriculture, under the present scenario appears to be an alternative. It is based on sound principles of soil biotechnology and microbiology. Indeed microscopic doses of few of the preparation have shown profound effect on growth, metabolism, crop yield and quality. It is interesting that these practices do not require sophisticated facilities and most of them can be created at farm itself by simple training.

9. Integrated intensive farming system (IIFS)

IIFS involves intensive use of farm resources. To be ecologically sustainable, such intensification should be based on techniques which are knowledge intensive and which replace to the extent possible, market purchased chemical inputs with farm grown biological inputs.

ADVANTAGES OF ORGANIC FARMING

- Organic culture contributes in the environment's protection maintaining the cleanliness of aquatic resources and promoting the biodiversity and the viability of countryside, seeing that quantities of chemical fertilizers and pesticides that, as we know, harm and influence considerably the harmony of environment, are not demanded.
- Organic farming produces the same crop variants as those produced via conventional farming methods, but incurs 50% lower expenditure on fertilizer and energy, and retains 40% more topsoil.

- The economics of organic farming is characterized by increasing profits via reduced water use, nutrient-contamination by pesticides, reduced soil erosion and carbon emissions and increased biodiversity.
- This type of farming effectively addresses soil management. Even damaged soil, subject to erosion and salinity, are able to feed on micro-nutrients via crop rotation, inter-cropping techniques and the extensive use of green manure.
- Farming the organic way enables farmers to get rid of irksome weeds without the use of any mechanical and chemical applications. Practices such as hand-weeding and soil enhancement with mulch, corn gluten meal, garlic and clove oil, table salt and borax not only get rid of weeds and insects, but also guarantee crop quality.
- The use of green pesticides such as neem, compost tea and spinosad is environmentally friendly and non-toxic. These pesticides help in identifying and removing diseased and dying plants in time and subsequently, increasing crop defense systems.
- Organic products protect the consumer's health, since they are products produced without agrochemical and other poisons that have negative effect in the human organism.
- Organic products have high alimentary value but at the same time an exceptional flavour. They are foods with big content in natural sugars, metals, micronutrients, vitamins, antioxidant cells and have small content in water (Wothington, 2007).
- Finally, supporters of organic foods claim that the reduced amounts of pesticides and chemicals in organic foods protect children and can prevent cancer, allergies, birth defects, nerve damage, genetic mutation and asthma (Lipson, 2001).

ORGANIC FARMING DISADVANTAGES

- In 1998, increased risk of *E. coli* infection via consumption of organic food rather than non-organic food was publicized by Dennis Avery of the Hudson Institute.
- A 2008 survey and study conducted by the UN Environmental Program concluded that organic methods of farming result

in small yields even in developing areas, compared to conventional farming techniques.

- The Father of the Modern Green Revolution, Norman Borlaug, argues that while organic farming practices are capable of catering to the demands of a very small consumer fraction, the expanding cropland is dramatically destroying world ecosystems.
- Research conducted by the Danish Environmental Protection Agency revealed that organic farms producing potatoes, seed grass and sugar beet are barely able to produce half of the total output churned out from conventional farming practices.
- Organic agriculture is hardly able to address or combat global climate change. Though regenerative organic farming practices are recognized as effective strategies for reducing CO_2 emissions to an extent, the impact is not dramatic.
- Finally, crop rotation prevents organic produce from being produced all year long, which means that unlike conventionally grown produce, organic foods may not be available during any given season.

Nevertheless, despite the various advantages and disadvantages of organic farming, farms where organic methods for cropping are experimented with, display more profitability than conventional farms. Organic agriculture is definitely more sustainable in the long-term, improving soil fertility and terrain drought resistance greatly. These farming practices completely waive off external costs, incurred due to investment in chemical pesticides and nutrient runoff, and a number of health issues that result from agro-chemical residue.

MANAGEMENT PRACTICES FOR ORGANIC FARMING FOR SUSTANABILIY

1. Organic manures in organic farming

Organic manure in a broad sense includes composts from rural and urban wastes, crop residues, agro industrial bio wastes and green manures, apart from the commonly used FYM. Organic manure improves soil physical condition including soil porosity and water holding capacity and microbial environment, replenishes essential micronutrients in soil, increases the utilization efficiency of applied fertilizers and favours micronutrient availability to the plant. Organic manure is of paramount importance not only in augmenting the crop production but also for making the agriculture sustainable as an ecofriendly means of soil health management (Mukherjee, 2008). It is well established that FYM plays an additional role than its capacity to contribute NPK. Unlike chemical fertilizers that supply only the major nutrients, FYM is a store house of several plant nutrients and acts as a good soil conditioner. Content of NPK in different kind of refuse, manure, oilcakes and slaughter house waste (Table 1 and 2) are useful in various crop sequences and help to restore soil fertility. In tropic and subtropical climate application of rice husk (@ 9 t/ha) serve as a source of Si nutrition which help in good growth of cereals (Mukherjee and Sen, 2005). Green manuring has recently been under practice by our farmers for decades. Estimates suggest that a 40-50 days old green manure crop can supply up to 80-100 kg. N/ha. Even if half of this N is crop utilizable, a green manure crop can be a substitute to 50-60 kg. fertilizer N/ha. Some of the potential green manuring legumes are dhanicha, sunhemp, cowpea, mung, bean, guar and berseem etc. dhanicha, sunhemp, mung bean and guar grown during *kharif* season as green manure crops have been reported to contribute 8-21 tones of green matter and 42-95 kg. of N/ha. Similarly, khesari, cowpea and berseem grown during rabi season can contribute 12-29 tons of green matter and 67-68 kg of N/ha.

TABLE 1

Average Nutrient Content of Bulky Manure

Manure	**Percent Content**		
	N_2	P_2O_5	K_2O
1. Animal Refuse:			
Cattle dung, fresh	0.3-0.4	0.1-0.2	0.1-0.3
Horse dung, fresh	0.4-0.5	0.3-0.4	0.3-0.4
Sheep dung, fresh	0.5-0.7	0.4-0.6	0.3-1.0
Night soil, fresh	1.0-1.6	0.8-1.2	0.2-0.6
Poultry manure, fresh	1.0-1.8	1.4-1.8	0.8-0.9
2. Wood Ashes:			
Ash, household	0.5-1.9	1.6-4.2	2.3-12.0
Ash, wood	0.1-0.2	0.8-5.9	1.5-36.0

Contd...

Contd...

Manure	Percent Content		
	N_2	P_2O_5	K_2O
3. Farm Factory and Habitation :			
Rural compost, dry	0.5-1.0	0.4-0.8	0.8-1.2
Urban compost, dry	0.7-2.0	0.9-3.0	1.0-2.0
FYM, dry	0.4-1.5	0.3-0.9	0.3-1.9
4. Plant Residues :			
Rice hulls	0.3-3.5	0.2-0.5	0.3-0.5
Groundnut husks	1.6-1.8	0.3-0.5	1.1-1.7
straw and stalks: Banana, dry	0.61	0.12	1.0
Maize	0.42	1.57	1.65
Paddy	0.36	0.08	0.71
Wheat	0.53	0.10	1.10
5. Green Manure, fresh			
Cowpea	0.71	0.15	0.58
Green gram	0.72	0.18	0.53
Sunhemp	0.75	0.12	0.51

Benefits of soil organic matter

- It serves as a slow release of nitrogen, phosphorus and sulphur for plant nutrition and microbial growth (Mukherjee and Sen, 2004).
- It posses considerable water holding capacity and thereby helps to maintain the water regime of the soil.
- It acts as a buffer against pH of the soil.
- Its dark color contributes to absorption of energy from the sun and heating of the soil.
- It acts as "cement" for holding clay and silt particles together, thus contributing to the crumb structure of the soil and resistance against soil erosion.
- It binds micro-nutrients metal ion in the soil that otherwise might be leached out of the surface soil.

TABLE 2

Average Nutrient Content of Major Oil Cakes and Slaughter House Wastes

Manure	Percent Content (%)		
	N_2	P_2O_5	K_2O
1. Oil Cake:			
a. Non-edible type:			
Castor Cake	4.3	1.8	1.3
Cotton Seed Cake	3.9	1.8	1.6
Mahua Cake	2.5	0.8	1.8
b. Edible Oil Cakes:			
Coconut Cake	3.0	1.9	1.8
Groundnut Cake	7.3	1.5	1.3
Cotton Seed Cake	6.4	2.9	2.2
Rape Seed Cake	5.2	1.8	1.2
Sesame Cake	6.2	2.0	1.2
2. Slaughter House Wastes:			
Meat Meal	10.5	2.5	-
Blood Meal	1.0-12.0	1.0-2.0	-
Fish Meal	4.0-10.0	3.0-9.0	0.3-1.5

2. Vermicomposting

Vermicomposting is an effective means of composting the decomposable organic wastes using earthworms naturally present in the soil. Vermicomposting is a mixture of worm casts enriched with macro and micronutrients (N, P, K, Mn., Fe, Mo, B, Cu and Zn.), some growth regulating substances such as gibberellins and auxins) and useful micro flora (*Azospirillum, Actinomycetes* and *Phosphobacillus*) etc. The nutrient level of vermicompost (1-1.5%N, 0.6-0.8% P and 1.2- 1.5 K) is higher than any other compost (Samanta, 2009). From the available information it is well documented that earthworms can consume all types of organic matter and convert them into available form of nutrients. Vermicompost improves the physical and biological condition of soil, improves soil fertility and pulverizes it through their churning and turning action in addition to contributing plant nutrients, improves aeration and water holding capacity (Mukherjee, 2009). It is reported that soils with casts contain 5 times nitrogen, 7 times phosphorus, 11 times potash, 2 times magnesium and 7-8 times actinomyces more than in soils without earth worm casts. It being a natural means of soil fertility management fits well into integrated plant nutrient management strategy for sustainable agriculture. Though more than 12 species are being used for vermicomposting across the globe the important species used for vermicomposting in India are *Eisenia foetida, Eudrilus Eugenia, Lampito mauritic, Pherionyx excavatus, Octochaetona serrata* etc. Earthworms can be divided into 2 groups. Epigeic (surface feeder) group of earth worms are very important in vermicomposting. Eg. *Eisenia foetida, Eudrilus eugenia, Pherionyx excavatus*. Second group include Epianecic group of earthworms are feeders on leaf litter in the upper layer of soil surface. Procurement of vermicompost is possible by preparing a vermibe and their size depends on the quantity and nature of the organic wastes to be composted.

3. Exploitation of biological nitrogen fixation

Atmosphere containing as much as 78% nitrogen can be a potential source of this essential nutrient to soil utilizing the nature's own mechanism of biological nitrogen fixation. It not only offers an economic and ecofriendly source of nutrient supply, moreover evidences show that nitrogen so derived is less prone to loss than fertilizer nitrogen and as a long term effect it builds up a reserve of readily mineralizable organic nitrogen. Thus it plays an important role in sustainable agriculture, reduces the requirement of externally applied nitrogen fertilizer and favourably influences the soil microenvironment. BNF is a natural system of biological mobilization of atmospheric nitrogen which can be easily available to and utilized by plants mediated by microorganisms like some eubacteria, cyanobacteria, actinomyces and a few archaebacteria which are commonly called diazotrophs (Roger, 1987). They have the ability to convert atmospheric elemental nitrogen (N_2) to ammonia (NH_3) with the help of nitrogenase enzyme system. Legume crop are mainly responsible for BNF (Singh, 2003) and fixation amount varies from crop to crop (Table 3). The efficacy of true bacteria like *Rhizobium, Azotobacter,Azospirillum* and cyanobacteria like *Anabaena,Aulosira, Nostoc, Plectonema* and *Tolypothrix* etc. have already been well established (Mukherjee, 2008 b) under field conditions and such microbial inoculants have gained wide acceptance due to their high nitrogen fixing potentiality (Table 4).

TABLE 3

The Range of N Fixed/ha. by Different Legumes

S.N.	Crop	Nitrogen fixed (kg/ha.)
1	Clover	100-150
2	Cowpea	80-85
3	Alfalfa	100-300
4	Faba Bean	240-320
5	Lentil	90-100
6	Lupins	150-200
7	Groundnut	50-60
8	Soybean	60-80
9	Mungbean	50-55
10	Pasture Legumes	100-400

TABLE 4

Method and Rate of Application of Rhizobium Culture

Crops	Method of application	Quantity of required /ha	Time of application
Chickpea, Peas, Beans, Groundnut, Soybean	Seed treatment	1.0 - 2.0 kg	Sowing time
Lentil, Lucern, Berseem, Green gram, Black gram, Cowpea, Pigeon pea, Guar	Seed treatment	0.4 - 0.6 kg	Sowing time
Sunhemp	Seed treatment	0.8 - 1.0 kg	Sowing time

(4) Utilization of green manuring crops in organic farming

Green manuring is a practice of ploughing or turning undecomposed green plant materials into the soil for improving the physical condition of soil or for adding nitrogen where the green manure crop is legume (Cheema *et al.*, 1997). The process of green manure is of two types :

(i) Green Manuring In-situ : A practice of ploughing or turning into the soil of undecomposed green manure crop in the same field where the crop is grown eg. sunhemp, sesbania etc.

(ii) Green Leaf Manuring : This refers to turning of green leaves and tender green twigs collected from shrubs and trees grown on bunds, waste lands and hereby forest areas. The common shrubs and trees useful for this purpose are glyricidia, sesbania etc.

Advantages of Green Manuring

- It adds organic matter to the soil
- The green manure crop return to the upper top soil, plant nutrients taken up by the crop from deeper layers
- It improves the structure of soil and other physical properties
- It facilitates penetration of rain water, thus decreasing run-off and erosion
- The green manuring crops hold plant nutrient that would otherwise be lost by leaching
- Leguminous plants add nitrogen to the soil
- It increases the availability of certain plant nutrients like phosphorus, calcium, potassium, magnesium and iron

The farmers in many parts of the hills of North Eastern Himalaya are also using wild plants such as Titepati (*Artemisia vulgaris*) and Siris (*Albizzia lebbeck*). Others are legume crops like Daincha, sun-hemp, berseem, cowpea, lentil etc. The use composition of some wild plants is given in the table 5.

TABLE 5

Wild Plants Used for Green Manuring and Their Nutrient Content

Wild Plants	Nutrient Conten (%)		
	N_2	P_2O_5	K_2O
Titepati (*Artemisia vulgaris*)	2.40	0.42	4.90
Asuro (*Adhatoda vasica*)	4.30	0.88	4.49
Taramandal (*Helianthus annus*)	4.90	0.87	5.23
Banmara (*Eupatorium adenophorum*)	2.35	0.71	-

(5) Cultural recommendations

(a) Minimum tillage

The conventional tillage operation has disastrous effect on soil erosion causing greater loss of nutrients (Mukherjee, 2008). This has led to necessity of stubble mulching. In the stubble mulching the soil is protected by the crop residues left on the soil surface during fallow periods. This has now become well established that in stead of frequent tillage operations minimum tillage is more useful not only because it offers cost effectiveness but also contributes to conservation of soil and moisture. Available information reveals that it gives as good as or even better yield than the conventional tillage practices. It aims at reducing tillage to the minimum, necessary for ensuring a good seed bed, rapid germination, satisfactory crop stand and favorable growing conditions.

(b) Selection of crops for crop rotation, inter cropping and companion cropping

Crop rotation has an important role to play in organic farming. Judicious selection of crops in crop rotation helps in efficient utilization of plant nutrient from different depths of soil. Study in the Darjeeling hills revealed that farmers practicing crop rotation and intercropping to improve their soil fertility and enhance soil productivity per unit area (Mukherjee and Moktan, 2008). The practice of inter cropping also helps in minimizing the risk of crop failure due to uncertain rainfall and infestation of pest and diseases. Available information reveals that in case of inter cropping of sorghum and red gram at 2:1 row ratio it resulted in 70% higher yield over the individual cropping of both the crops. Besides it also reduced the wilt diseases of red gram. Another advantage was that nitrogen fixed by the red gram could be utilized by the sorghum crop. Another trial of companion cropping comprising of three rows of gram followed by one row of linseed revealed that incidence of pod borer in gram as well as fly in linseed was remarkably reduced and yield also increased by about 14% in this inter cropping over the sole cropping of both the crops.

6. Integrated pest management (IPM) programmes for sustainable agriculture

IPM is a procedure to manage pest populations by harmonizing control methods such as natural enemies, pesticides and cultural practices. The purpose of IPM is not eradication or removal of the pest, but management of pest populations so that economic damage and harmful environmental side effects are minimized. Modifications of crop environment to make it unfavourable for infestation of pests, use of resistant cultivars and transgenic along with a judicious and need based use of preferably safer chemicals are the major components of IPM. Biocontrol is an integral component of integrated pest management strategy. It employs natural mechanism of pest suppression in which the natural agents like predators, parasites and microbial agents suppress the pest population. If the natural enemies can be effectively conserved, the need of other control will automatically reduce.

(1) Biological pest control

In Korea, by using inorganically method of pest control in the rice is becoming popular nowadays. Current biological control research is mainly concerned with the brown plant hopper (BPH) on rice. Biological control of the BPH is best by the spider and fungal pathogens. A number of leaf hopper and BPH are controlled by using different types of spiders in Koreas was reported by Choi and Lee (1990) and their research report is tabulated in Table 6.

TABLE 6

Number of Leaf Hopper and Plant Hopper Preyed per Spider per Day

Pests	*Pitrata subpiraticus*			*Pachygnatha clerki*			*Ghathonarium dentatum*		
	Male	Female	Mean	Male	Female	Mean	Male	Female	Mean
Nilaparvata lugens	6.8	3.7	5.3	2.2	1.7	2.0	1.2	1.0	1.1
Sogatella fercifera	3.8	4.0	4.0	2.8	1.9	2.4	1.2	0.9	1.1
Laodephax striatella	11.9	6.5	6.5	2.1	1.8	2.0	2.6	0.9	1.3
Nephotellix cinticeps	5.8	3.1	3.1	2.6	2.1	2.4	2.5	1.9	2.2

Source: Choi and Lee, 1990

(2) Biological weed control

One aspect of biological control of weeds is direct use of natural enemies to reduce weed populations (Mukherjee and Singh, 2004).

They are usually plant pathogens but may be insects or various herbicides, for eg. tadpole, ducks etc. are used for weed control in India (Mukherjee, 2005 a), as in the grass carp in Indonesia and the apple snails in Taiwan.

(3) Bio-pesticides for insect-pest and disease management

The growing public awareness of the chemical insecticides used to control insect-pests has urgent the scientists to develop new bio-pesticides as an alternative (Verma, 2009). Bio-pesticides are distinguished from conventional chemical pesticides by their non-toxic action. They include the following:

(a) Microbial Pesticides : Microbial pesticides are naturally occuring organisms which include bacteria, fungi, protozoa or viruses eg. BT (*Bacillus thuringiensis*), Baculoviruses etc. (Mukherjee, 2007 a). Some of the successful examples to control crop pests are as follows:

(a) Entomogenus fungi eg. *Metarhizium anisopliae* and *Isaria sinclavii* were tested and applied to control sugarcane grass cicada.

(b) *Verticillum lecani* was infective to corn aphids

(c) *Entemophthora* spp. were reported to be highly pathogenic to the tiger moth, *Creatonotos gangis* and the green leaf hopper (Roger, 1987)

Baculoviruses are promising agent for the control of insects of order Lepidoptera (Butterflieds and moths), Hymenoptera (Sawflies) and Coleoptera (Beetles).

(b) Biochemical Pesticides : Biochemical pesticides from insect and plants which can be used to modify insects' behavior and physiology and even affect insect control. These include semio-chemicals, hormones, natural plant regulators, enzymes etc. Sex pheromones are one kind of semio-chemicals that attract the pests and controlled by mechanical damage. About 1000 insect pheromones have been isolated and identified and synthesized in Taiwan (Dhaliwal and Arora, 1998)). Among them, botanical pesticides investigated, neem has justifistically received the maximum attention during the last two decades. More than 300 species of insects have been reported to be affected by the neem components (Singh, 1993).

Advantages of bio-pesticides

- Lack of residues and pollutants in the soil

- High level of safety to human and non-target organisms
- Low livelihood of pest resistance
- Environmentally safe
- The are selective, biodegradable, ecological and renewable alternative for the use of Integrated Pest Management (IPM) programs

PESTS AND DISEASE CONTROLLING MEASURES THROUGH INDIGENOUS ORGANIC BASED TECHNOLOGY

In recent times farmers in the hilly region of Darjeeling district and north eastern himalaya range, have been frustrated with continuous chemical sprays yielding lesser outputs (Mukherjee and Moktan, 2008). Both low yields and increased cost of production have disturbed their minds and lives. Human and soil health have also been affected severely. The application of chemicals for pest control leads to food poisoning, soil, water and environmental pollution. These chemicals create ecological imbalance and allow insect-pest to develop resistance. In this context, integrated pest management through eco-friendly practices would play an important role to keep a check on the problems and improve the livelihood of the farmers.

Control of termites

(a) Pour sugar and flour around old tree trunk to attract common ant. This practice helps to maintain required population build up of ants in the area which feed on termites.

(b) Application of a paste on the main trunk of fruit trees with a mixture of mustard oil (1 litre) + turmeric powder (200 g) attract ants which feed on termites.

(c) Chopped oak leaves filled in gunny bags are kept in irrigation channel. The exudates from these leaves are reported to kill termite.

(d) Use of kerosene is also common practice against control of termites in the field.

(e) Mixture of 10 kg of ash with a 1 kg of sugar is effective to control termites.

Control of aphids, thrips, white flies and other pests

(a) Application of wood or cow dung ash on foliage of vegetables, eg. Chilli, onion and cucurbits to prevent insects such as thrips and aphids is a common practice.

(b) 4 kg tobacco leaves and twigs are boiled in 40 litres of water for 40 min. After cooling, one kg soap powder is mixed and soluition is diluted 7-8 times and sprayed to control aphids and white flies in kinnow and other citrus plants.

(c) Smoke created by putting sulphur over hot cow dung has been reported to kill/repel aphids and other sucking pests of mango.

(d) Maize seeds are soaked in cow urine for 10-12 hr. before sowing for resistance against insect pests.

(e) Rice seedlings raised from seed treated with extract of neem kernel are resistant to leaf hopper.

(f) In rice, spraying a solution of 4 litres of cow urine and 10 g asafoetida in 10 litres of water repel aphids and jassids.

(g) For prevention of infestation of shoot borer common salt is mixed with soil near the collar region of tree.

(h) In case of insect holes made by shoot borer and bark eaters in fruit plants, jaggery is placed in the holes to attract other predators to feed on the insect present in the hole. Similarly, the practices of pouring kerosene or petrol in holes and blocking holes with mud or cow dung are also common in citrus plants.

(i) In Arunachal Pradesh, farmers believe that the use of cow dung slurry can protect crops from aphid attack and act as repellent to cutting and biting insects and animals. Cattle urine has successfully used against thrips, mites, aphids and caterpillars. Application of extracts prepared from the parts sisnu and fruits of timbur plants are used to control many kinds of chewing, biting and cutting insects.

(j) Adi tribes of Basar use five wooden / bamboo sticks randomly in the rice field during flowering stage and place dead frog/dry salted fish/crabs on the stick to attack adult Gandhi bugs.

(k) Chilli garlic extract: Three kilos of green chilli are made into a paste and soaked in ten litres of water overnight. Grind 500 gms. of peeled garlic and soaked in 250 ml of kerosene and kept overnight. Next day these two extracts are filtered separately. 80 gms. of detergent less soap is diluted in one litre of water. And the three solutions are mixed and left for four hours. Then dilute the extract adding 70 litres of water and spray can be made when the climate is cool.

(l) An extract of tobacco waste with 250 g of soap solution in 200 litres of water as spray can control stem borer.

(m) A solution prepared from neem leaf paste in water (10 kg in 2 litres) is effective to control leaf folder in rice.

(n) A solution prepared from 100 g tulsi leaves in 1 litre of water with 1 ml. soap solution can be used for effective control of aphids, army worm, red cotton bug, mosquito etc. is used to control cotton semilooper, mites, green leaf hopper, aphid etc.

(o) Trap crops underline the need for mixed-cropping in the fields. Ladies finger attract a large number of pests. Likewise crysanthamum, maize, red gram, mustard and cowpea are grown in farmers fields to protect their main crop. This is one of the easiest methods of controlling the pests.

Control of stored grain pests

(a) Green gram can be kept free from any pest infestation by treating with 1% neem leaf powder.

(b) There is a common practice of storing food grains in wheat or barley straw to prevent storage pest damage.

(c) A layer of sand (5-7 cm.) is effective to protect pulses from beetle attack.

(d) Pulses mixed with little cow dung powder in earthen pots can be stored for long time.

(e) Millet, chickpea, green gram and other pulses can be stored well after thoroughly treated with mustard oil.

(f) Garlic leaves can be used for safe storage of food grains.

Plant disease control

There are many non-chemical methods for managing disease. Plant pathogen can be avoided by proper selection of the field, by selection of disease escaping varieties and other planting materials by manipulating sowing time, planting density etc. Traditional systems of crop protection that farmers have learnt from their long association with the land, its flora and fauna were based on eco-friendly systems of suitable cultural practices, judicious rotation of crops and knowledge of pests and diseases and their life-cycles.

(a) In rice, spray with a solution of cow dung prepared by mixing 3 kg cow dung in 3 liters of water is found to be effective against the control of rice blast and bacterial blight.

(b) In case of bunchy top disease in chillies, dusting of ash, spray of sour butter milk, spray of liquid waste of tanned leather and spray of cow/goat urine has been recorded in the tribal areas.

(c) Some farmers used to apply fresh cow dung near the collar region of chilli plant to control fungal diseases, viz. damping off and dieback.

(d) In case of soil-borne diseases, root rot, collar rot, the castor cake, karanj cake or neem cake were used as control measures.

(e) The milk solution (1 litre milk in 9 litres of water) is effective to control powdery mildew and viral diseases.

(f) 20 kg of *Casuarina equisetifolia* leaves are boiled in water for 20 min. After cooling, the solution is filtered. Then the extract is diluted with water and can be given to control some bacterial and fungal diseases.

(g) Wheat seeds are soaked in milk before sowing to avoid rust diseases.

(h) A solution is prepared from 2 kg fresh leaves of papaya in 3-4 litres of water and kept it overnight. After filtration, this is diluted with 50-60 litres of water and 250 ml. of soap solution. This is effective to control brown spot disease of rice.

(i) Cultivation of marigold followed by solanaceous vegetable crops is effective to control bacterial diseases and nematodes.

(j) Leaves of khair can be put into water channel to control brown spot disease of rice.

APPROVED MATERIALS AND PRACTICES FOR ORGANIC FARMING

The following **insect pest control materials and practices are**

approved for use in organic crop production. Some approved materials have certain restrictions regarding their use in organic crop production. Always carefully read the label and any other documentation. Apply all materials with awareness and care.

Approved materials	Avoid this things
Bacillus thuringiensis	Liquid forms containing xylene are prohibited.
Beneficial insects	
Boric acid	Cannot be used on edible plant parts.
Codling moth granulosis virus	
Cryolite or sodium fluoaluminate	The mined material from Greenland is permitted.
Diatomaceous earth	Use a dust mask when applying to prevent lung irritation.
Dormant oils	Use only on woody plants as a dormant spray.
Garlic	
Herbal preparations	May not be extracted with synthetic solvents.
Insect extracts	
Nematodes	
Pheromones	
Pyrethrums	Naturally occurring forms are allowed. The pyrethrums are highly unstable in the presence of air, light and moisture. They have low mammalian toxicity and can cause dermatitis in humans. Use with caution.
Rotenone	Use with caution. Rotenone is highly toxic to fish. Its persistence in the soil is unknown though it loses its effectiveness within one week. Should not be used on crops nearing harvest time. The active compounds, rotenoids, are present in a variety of legumes including soybeans.

Contd...

Contd...

Approved materials	Avoid this things
Ryania	Use with caution.The toxicological properties of ryania are largely unknown.
Sabadilla	Use with caution.
Shells, ground	Oyster, clam, lobster and crab
Soaps	
Sulfur, elemental	
Summer oils	May be used on woody plants only, carrot and/or weed oils are prohibited.
Trapping substances	
Tree seals	May be petroleum based but may not contain synthetic chemicals or fungicides.
Virus sprays	

The following **weed control materials and practices are approved** for use in organic crop production.Some approved materials have certain restrictions regarding their use in organic crop production. Always carefully read the label and any other documentation. Apply all materials with awareness and care.

- Laming
- Grazing
- Herbicidal soaps
- Mechanical and cultural controls
- Mulches of organic materials
- Plastics for mulch, row covers and solarization
- Weeder geese

The following **disease control materials and practices are approved** for use in organic crop production. Some approved materials have certain restrictions regarding their use in organic crop production. Always carefully read the label and any other documentation. Apply all materials with awareness and care.

Antibiotics	Naturally derived antibiotics are permitted for disease control.
Bordeaux mixture	Use with caution. Excessive use of bordeaux may cause buildup of copper in the soil and limit the continued use.
Copper hydroxide	
Copper sulfate	Use with caution.Excessive use of bordeaux may cause buildup of copper in the soil and limit the continued use.
Dormant oils	Use only on woody plants as a dormant spray.
Hydrated lime	Foliar application as a fungicide only. May not be used as a liming material.
Hydrogen peroxide	
Lime sulfur	Foliar application as a fungicide only.
Soil pasteurization	
Sulfur, elemental	
Tree seals	May be petroleum based but may not contain synthetic chemicals or fungicides.

The following **vertebrate pest control materials and practices are approved** for use in organic crop production. Some approved materials have certain restrictions regarding their use in organic crop production. Always carefully read the label and any other documentation. Apply all materials with awareness and care.

Deer and rabbit repellents	Acceptable if derived from a natural source.
Predators	Cats, hawks, coyotes, airborne projectiles.
Rodent traps	
Strychnine	Underground use only.
Synthetic vitamin baits	

Prohibited materials and practices

Insect pest control materials and practices that are prohibited for use in organic crop production include, but are not limited to:

- Abamectin or Avermectin
- Carbamates
- Chlorinated hydrocarbons
- Dimethyl sulfoxide
- Methyl bromide
- Methyl sulfoxide
- Moth balls/crystals
- Nicotine
- Organophosphates
- Plant protectants, synthetic
- Pyrethroids, synthetic

Weed control materials and practices that are prohibited for use in organic crop production include but are not limited to(Mukherjee, 2007):

- Carrot oil
- Synthetic growth regulators
- Herbicides
- Weed oils

Disease control materials and practices that are prohibited for use in organic crop production include but are not limited to:

Soil fumigants	Synthetic fungicides, fumigants, sterilants and bactericides

Vertebrate pest control materials and practices that are prohibited for use in organic crop production include but are not limited to:

- Anticoagulant rodent baits

ORGANIC FARMING IN CONTEXT OF INDIA FOR SUSTAINABLE ISSUE

India had historically practiced organic farming. Sir Albert Howard, who was sent to India as an Imperial Economic Botanist to work at the Indian Agriculture Research Institute in the year 1905 to improve Indian agriculture, soon concluded that agriculture, as practiced by the India peasants, was rooted on the sound principles of sustainability. "Howard" who was elected President of Indian Science Congress in 1926, observed that agriculture research should not be misused to make the farmer exploit the soil reserves but to teach him the know-how to transfer capital in the shape of soil fertility and the reserves of his livestock to his profit and loss account. Organic farming is gaining growing importance in the agriculture sector of a number of countries, irrespective of their stage of development. In several developed countries, organic agriculture has come to represent a significant portion of the food system (10% in Australia, 7.8% in Switzeriand) and many others are experiencing growth rates that exceed 20% annually (e.g. USA, France, Japan, Singapore). Some of the developing countries have small domestic organic markets (e.g. Egypt) and a few have begun to seize the Iucrative export opportunities presented by organic agriculture (e.g. exports of Mexican coffee, Ugandan cotton.).

A comprehensive policy on organic farming has been proposed by the Ministry of Agriculture. The Indian Govt. promotes organic farming to emphasize the need to reduce the use of harmful chemicals on the farm. The government identifies progressive farmers to participate in organic farming training, and to help them to form "Organic Farming Association" at the village level. Three types of organic farmers in India:

- Farmers who follow the old pattern of indigenous farming practice;
- Farmers who are practicing "biodynamic agriculture" or "natural farming" on their own small- and medium-sized lands; and
- Private companies engaged in large-scale organic farming for export.

India produces primary organic products (coffee, tea, spices, fruits, vegetables, cereals, as well as honey and cotton), and processed foods are limited. Few of the medicinal plants ex. *Swertia chirayita* and *Valeriana jatamansi* etc. has high value in pharmaceutical company if grow organically (Mukherjee, 2008 a). Organic husbandry, poultry, and fishery do not exist. Domestic organic markets and consumer awareness are underdeveloped in India, but interest is growing. On the domestic market, organic food is usually sold directly from farmers or through specialized shops and restaurants. At present, organic products receive a price premium of 20-30 per cent over conventional products. India is an exporting country, mainly to Europe and USA, and does not import any organic products.

In addition, there is a need to earnestly orchestrate an organic farming revolution in which following issues need to be given stress and close attention: -

1. Generating awareness among the farmers to change their mindset.
2. Guiding and helping the farmers in the proper implementation of organic farming techniques.
3. Making proper arrangement for marketing the organic fertilizer and organic pesticide.
4. Helping the policy maker and administration to come forward with sound policies to help the farmers.
5. Increase agro-bio diversity through mix-cropping, conserving water resources and increasing genetic diversity.

External certification bodies introduced inspection and certification program in 1989 (FAO 2002). The Indian Organic Standards were modeled on the IFOAM Basic Standards and the seal "India Organic" was established. In October 2001, the export of organic products was brought under government regulation, while import and the domestic market were not. India's first local organic certification body, "Indocert," was founded in March 2002. Indocert's aim is to provide reliable and affordable organic inspection and certification service to farmers, processors, and input dealers. It provides certification for the domestic and export market.

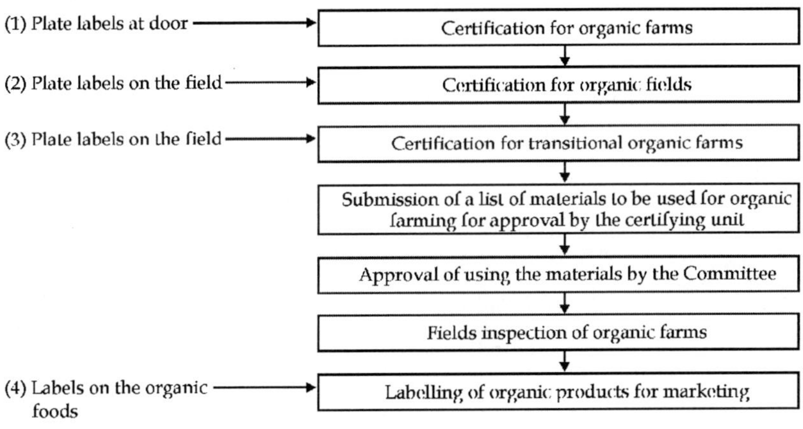

Fig. 1 : Flow Chart of Organic Food Certification and Labeling in India

To be certified organic in India, products must be grown and manufactured in a manner that adheres to standards set by the country (Fig. 1). A few of the apex certification agencies around the world are:

India	NPOP, (National Program for Organic Production)
United States	National Organic Program (NOP) Standards, by USDA
Australia	NASAA Organic Standard
Britain	Organic Farmers and Growers Organic Standards and the Soil Association
Canada	Canada Gazette, Government of Canada
European Union	EU-Eco-regulation

CONCLUSION

Agriculture remains the key sector for the economic development of most developing countries, because for the development of any country should be self sufficient for food shelter and cotton. To make self sufficiency of the country, there has been intensive fertilizer used for the last four decades, which created several problems linking excessive fertilizer used with environment have been identified. Increased amount of nitrate in drinking water is due to excessive and improper use of nitrogen fertilizers, which is most important fertilizer related pollution issues. Nutrient enrichment, eutrophication of lakes and deterioration of surface water quality due to transportation of nutrients applied through fertilizer via leaching and/or runoff and sediment erosion are other problems. Contamination of soil by heavy metal through fertilizer such cadmium from phosphatic fertilizer which has been found to carcinogenic is also getting increasing attention of environments. The significance of fertilizers in polluting the air is being viewed with concern after the course that nitrous oxide originating from agricultural soil can damaged the ozone layer in the stratosphere.

Pesticide contamination of agro-chemical and horticultural crops have also been reported. Agriculture is not sustainable if its resource base declines, or if it has an adverse impact on the environment or if it leads to economic hardship for farmers especially for farmers with limited resources and landless tenant cultivators. To overcome such problems, organic farming receives the top priority in sustainable agriculture. There is a considerable scope for supplementing of renewable rsources such as bio-fertilizer and organic wastes for improving crop productivity and crop health. The use of FYM, compost is in practice but the use of bio-fertilizers, green manures and other waste has yet to become popular as a cost effective and eco-friendly nutrient supply. Experiments conducted on different legume crops grown under varying agro-ecological conditions proved the potentiality of bio-fertilizer and organic wastes as important source of plant nutrients. So, from the different aspect, it is clear that organic farming is practical proposition for sustainable agriculture if adequate attention is paid to this issue. There is urgent need to involve more and more scientists to identify the thrust area of research for the development of eco-friendly production technology.

REFERENCES

Asmed, S. 1993. Agriculture-Fertilizer Interference in Asia. Issue of Growth and Sustainability, Oxford and IBH Publishers, New Delhi, India. pp. 23-29

Bould, C. 1994. Diagnosis of Mineral Disorders in Plants, Vol. 1,2, 3 Principles. Chemical Publishing, New Delhi, India.p 64.

Bourn, D, Prescott J. 2002. A comparison of the nutritional value, sensory qualities, and food safety of organically and conventionally produced foods. Critical Reviews in Food Science and nutrition. pp. 42: 1-27.

Bourn, D., and Prescott, J. 2002. "A Comparison of the Nutritional Value, Sensory Qualities, and food Safety of Organically and Conventionally Grown Produced Foods." Critical Reviews in Food Science and Nutrition 42(1):1–34.

Cheema S.S., 1997. Agronomy (Theory and Digest), Kalyani Publishing, New Delhi, India.

Choi, K.M. and Lee M. II., 1990. Use of Natural Enemies to Control Agricultural Pests in Korea, FFTC Ext. Bull. 304: 10-20.

Dahama, A.K., 1997. Organic Farming for Sustainable Agriculture, Ashila Offset Printers, Daruagung, New Delhi, India. pp. 78-85

Dhaliwal, G.S. and Arora, R. 1998. Principles of insect pest managemtn. Ludhiana : Kalyani publishers. pp. 142-165.

Fisher, A. Ramirez, H.V. and Lozano, J. 1997. Suppression of jungalrice (*Echinochloa colona*) by irrigated rice cultivars in Latin America. *Agronomy Journal* 89 : 516-521.

Halweil, B. 2001. Organic Gold Rush. World Watch. (14): 22-32.

Hegde, P. 2009. Climate debate can't ignore small farmers. LEISA India. p. 20.

Hunter, B.T. 2002. Organic food goes mainstream. Consumers' Research Magazine. (8) : 85.

Kleiner, S.M. 1995. Should you opt for organic?. The Physician and Sports Medicine. (23) : 15-16.

Lipson, E.M. 2001. The Organic Foods Sourcebook. Chicago, IL: Contemporary Books; pp. 74-80

Long, C. Finding truly good food. Mother Earth News. 2002 (189) : 70-72.

Mukherjee, D. 2002. Food security coping with nutritional challenge ahead. *Agriculture Today* 8 : 30-32.

Mukherjee, D. 2003. Awareness programme for sustainable development in agriculture. *(In) Abstract International seminar on communication and sustainable development in agriculture,* January 7 – 9, Institute of Agricultural Sciences, Banaras Hindu University, Varanasi, p. 113.

Mukherjee, D. 2004. Kadia sureksha ke bartaman chunautia (in Hindi). *Surebhi Sandesh* p. 40.

Mukherjee, D. and Singh, R.P. 2004. The biological control of weed (A Review). *Agricultural Review* 25 (4) : 279-288.

Mukherjee, D. 2005. Eco-friendly tools for weed management in rice and wheat. In: *TRENDS IN ORGANIC FARMING IN INDIA.* (eds. Purohit, S.S. and Gehlot, D). Agrobios India, Jodhpur pp. 372-379.

Mukherjee, D and Sen, A. 2005. Influence of rice husk and fertility levels on the nutrient content of rice (*Oryza sativa* L.). *Agricultural Science Digest* 25 (2) : 151-152.

Mukherjee, D. 2005 (a). Eco-friendly tools for weed management. *Kisan world* 40 (10) : 51.

Mukherjee, D and Singh, R.P. 2006. Studies on phytotoxic effect of sulfonylurea herbicide on different rice cultivars. *Indian Journal of Weed Science* 38 (1&2) : 129-130.

Mukherjee, D. 2006. Weed control in rice through micro herbicidal approach. *International Journal of Agricultural Sciences* 2 (2) : 452-455

Mukherjee, D. 2007. Weed management strategy in rice -A Review. *Agricultural Review* 27 (4) : 247-257.

Mukherjee, D. 2007 (a). Effect of different organic source of nutrient on growth and yield of wheat (*Triticum aestivum* L.) in hilly region. International Symposium on Organic Farming and Renewable Sources of Energy for Sustainable Agriculture. November, 19-21, 2007, Maharana Pratap University of Agriculture & Technology, Udaipur. p. 33.

Mukherjee, D. 2008. Effect of tillage practices and fertility levels on the performance of wheat (*Triticum aestivum*) under mid hill condition of West Bengal. *Indian Journal of Agricultural Sciences* 78 (12) : 1038-1041.

Mukherjee, D. 2008 (a). Potential of herbals (medicinal plant) in context of India. MAPs Dew 4 (3) : 1-5.

Mukherjee, D. 2008 (b). Effect of different biofertilizer and organic source of nutrient along with chemical fertilizer on wheat under mid hill situation. *Indian Agriculturist* 52 (1&2) : 49-52.

Mukherjee, D and Moktan, M.W. 2008. Studies on land holding and crop diversification pattern in hill region of West Bengal. *Journal of crop and weed* 4 (1) : 41-45.

Mukherjee, D. 2009. Medicinal plant in Darjeeling hills. In : *KRISHI SANDESH* (eds. Rai, S., Moktan, M.W. and Ali, S). Miazik International Volunteer Center, Japan. pp. 118-121.

Roger, F.H., 1987. Importance of Bio-fertilizers in Intensive Cropping, Haryana Farming, Gundhi Bhawan, CCSHAU, Hisar, India. pp. 84-86.

Samanta, T. 2009. Use of vermicompost. *Indian Farming* (September) pp: 38-39.

Singh, C. 1993. Modern Techniques of Raising Field Crops, Oxford and IBh, New Delhi, pp. 148-160.

Singh, G. and Sekhon, H.S. 2002. Organic farming. Recent advances in Agronomy. *Indian Society of Agronomy*. pp. 167-189.

Singh, R.P., Singh, R.K. and Mukherjee, D. 2004. Effect of weed interference on efficiency of crop. *Agronomica* pp. 24-26.

Verma, R. 2009. Development of biopesticide. *Indian Farming*. (September). p. 2.

Worthington, V. 2007. Nutritional quality of organic versus conventional fruits, vegetables, and grains. The *Journal of Alternative and Complementary Medicine*. (7) : 161-170.

❑❑❑

6

Vermicompost : Role in Organic Farming

MOHIMA PRASAD BEHERA AND JOGESWAR SAHOO

Disposal of organic wastes from domestic agricultural and industrial sources has caused increasing environmental and economic problems. The waste are sewage, sludge, solids form waste water materials from food processing brewery industries, paper, cotton , cassava, mushroom and similar industries. Wastes from hotels, canteen, supermarkets, vegetables markets, poultry and pig industries, sheep, dairy, horse manure and agricultural wastes have great potential, if processed suitably, not only from reducing pollution but for recycling as a regular farm and horticulture input and can be used as an efficient input for organic or biological farming.

Efforts are being made, world wide, to evolve technologies for quick composting with minimum inputs and facilities, for processing these wastes. Though major thrust is on improving C : N ratio of the raw wastes and use of various decomposing cultures, use of earthworms at appropriate stage of decomposition is also gaining importance during last two or three decades.

Most organic wastes can be broken down by earthworms easily but many need to be partially decomposed before earthworms can take up the processing.

VERMICOMPOST

The cast of these earthworms is in the form of small granules like tea powder and is rich in humus, some undecomposed organic matter, micro organisms, minerals and nutrients. It also contains enzymes like protease, analyze, lipase, cellulase and chitinase. Vermicompost is recognized as the best organic manure today. Hence, *Eisenia foetida* and *Eudrilus euginea* are the best species of earthworm and useful in the waste disposal systems. Organic waste generated at cities; fruit and vegetable processing units, sugar factories, paper mills, can be converted into vermicompost. These wastes should be pretreated first by particular bacterial culture. This method of vermicomposting involves less energy, less labour and low costs compared to other kinds of mechanical composting methods.

It also keeps the environment clean. However, there is need of tireless efforts and creative programme to popularise the concept of vermicomposting. The end user i.e. the farmer ought to be educated properly in the various species of earthworms, their characteristics, food habitats reproduction rate and their role in agriculture. He ought to know how this scientific knowledge about earthworms be useful in improving the quality of soil and crop and increasing the organic productivity of his land.

ORGANIC AND AGRICULTURAL WASTES FOR COMMERCIAL VERMICOMPOSTING

Cattle dung, a bi-product of dairy and agriculture industry is a common material used for vermicomposting. Many organic wastes including animal, poultry, vegetable and farm wastes sewage sludge, town waste, sugarcane trash, etc. can be used for vermicomposting. Harvested crop bi-products, cotton waste and paper mill wastes have also been tried successfully. Banana stem peels, stems of crops, husks and wastes or discarded cattle feed can be used. Coconut coir, grass, husks etc. can be used as bedding material. Most of these wastes are bulky and transport costs are heavy for the same. Sizeable vermicomposting plants established near vegetable market yards city waste clumps, close to sugar, cotton, cassava mills, big dairy farms would be desirable, irrespective of the comparative costs of Vermicompost at present market rate.

SELECTION OF SITE

Low transport cost, easy availability of organic wastes throughout the year availability of cattle dung for mixing, sufficient open space and water sources are major requirements for starting the unit. A site on little elevation and little separated from human habitation should be preferred however in city farm houses it has been successfully raised. Consideration of market for Vermicompost in the close by area or selling through different organic farms or Co-operative Societies must be ensured prior to raising.

Suitable species of earthworms for vermicomposting

Different earthworm species have been identified suitable for waste decomposting of these the following four are most commonly used in sub-tropical and tropical countries.

(i) *Eisenia foetida* (Branding or tiger worm)

(ii) *Eudrilus eugeniae* (African night crawler)

(iii) *Perinox excavatus*

(iv) *Dendrobanea veneta*

Of these first three are highly suitable for tropical conditions, especially for central and southern India. Again *Eisenia foetida* and *Eudrilus eugeniae* are more common than the third species. Both these species are epigeic, surface feeder, ingest little or no soil, with heavy pigmentation, small to medium in size, responding rapidly to disturbances, relatively short lived, with rapid multiplication and cacoon stage, being slightly susceptible to predators because of surface-feeding habit and very easy for vermiculturing and handling. *Eisenia foetida* is very common as it is slightly temperature tolerant (15-30°C), tolerant to high moisture content of waste, rather a though worm though smaller in size. *Eudrilus eugeniae* as little large in size, very rapidly growing profile with high cacoon number, efficient in waste intake, decomposition and excretion. Under high temperature condition this worm should prove more suitable and very active except under low winter temperature. Both these worms are cultured commercially and are easily available.

OTHER FAMILIES OF EARTHWORM

Among the various families of earthworm Megtascolicid and Lumbricidae are of special importance. Lumbricidae earthworms are more useful in the progress of human being because they are able to adopt a new place in varied conditions. They were introduced to many parts of all temperate regions of the world by European countries. Eudrilidac is another family of earthworm which has been settled in Africa. *Pheretima elongata* which is found in India is of Megascalecid family. They burrow deep and are yellowish in colour.

Some important species of earthworms from Lumbricidae family are as follows :

(i) Lumbricus rubellus

This is deep burrower in nature and red brown in colour. Its length is 3 to 10 cm. *Lumbricus terestris—Allolobofora rosea, Allolobophora caliginosa, Alolobosphora longa, Allolobophora nocturna, Ȧllolobophara chlarotica. Eisenia foetida* is also from lumbricus family. These species are found in Europe and central Africa.

In south India many species are found - *Dichogaster offinis, Dichogaster bolavi, Dichogaster curgenis, Drawida bolivi, Lampito maruitii, Megascolex Konkanerisis, Poly pheretima elongata*, etc. Bouche (1971,1972, 1977 b) recognized three major morpho-ecological groups among European Lumbricidae. He distinguished the groups primarily on the basis of morphological characteristics that he considered to have functional significance on the basis of the soil horizon in which they were most commonly found. These were -

(i) Surface earthworms

(a) Epigees : Species that live above the mineral soil surface, typically in the litter layers of forest soils. Other related subgroups were those live under bark of trees, that feed on decomposition phloem, in cracks and crevices above ground, in temporary accumulations of organic matter.

(b) Endogees : Species that inhabit mineral soil horizons, feeding on soil more or less enriched with organic matter. Perel classified lumbricid earthworms into morpho-ecological categories.

(ii) Tunnel earth worm

Aneciques : Species that live in burrows in mineral soil layers, but come to the surface to feed on dead leaves which they drag into their burrows.

Difference between humus formers and humus feeders

The resistant materials like lignin and humic substances obtained after continued decomposition of plant and animal wastes i.e. organic wastes are the humus in the soil. Earthworms either form it or feed it.

Humus formers or surface earthworms	Humus feeders or earthworms forming tunnels
1. Simple intestine with a small typhosole that has no secondary folds	1. Complex intestine, essentially cylindrical but with a large involuted typhlosole, secondary fold the increase the surface area available for absorption
2. Purple or dark brown pigmentation	2. Little or no pigmentation
3. Fattened tail, closed epibolic or tanylobic prostomium which can be readly maneuvered to grasp plant debris from the ground surface.	3. No flattening of tail. Open epibolic prostomic with little independent maneuverability.
4. High mobility related to a rather complex pinnate arrangement of the longitudinal muscles.	4. Low mobility, related to arrangement of the longitudinal muscles in a more simple bundle like discompsotion.
5. Responds to UV light 64-69% of the lethal limit	5. Delayed response to physical stimuli- responds to UV light at 78-80% of the lethal limit
6 Gas exchange rates higher	6. Gas exchange rates lower

ROLE OF EARTHWORMS INFLUENCING THE PHYSICAL, CHEMICAL AND BIOLOGICAL PROPERTIES OF SOIL

Soil particles are cemented by calcium humate derived from the interaction of ingested organic matter and calcium excreted by the calciferous glands (Meyer, 1943, Arthur, 1965). They noted that the bacterial population of the earthworm gut is some 100 times that of surrounding soil and many bacteria produce polysaccharide molecules that are known to bind quartz and clay particles.

Earthworms that inhabit burrows provide the pathway for the movement of surface water and large particles from the surface to deeper layers and ready access for plant roots to penetrate the soil and also influence on pore space and thus on water infiltration and aeration in the soil. Earthworms affect the pore space in soil by burrowing and by depositing their casts (Porosity of soil increases by 75 to 100 per cent due to earthworm (Hoeksema and Jongerius 1959). Satchell (1967) estimated that two third of the airfield porespace of soil may be earthworm burrows. Carter et al. (1982) found that water entereci the soil at 2 to 10 times the rate in soil with earthworms than it did in soil lacking earthworms. These are good effects of earthworms burrows on aerations on infiltration and drainage.

Decrease in bulk density and increase in porosity of soil was observed due to the application of FYM / organic matter (Biswas and Khosla (1971), Prasad and Singh (1980). Rayuar (1984) reported that application of FYM significantly increased the water holding capacity and porosity in all textured soil. Hapse et al (1993) reported that application of vermicompost decreased the bulk density and increased the porosity of the soil as compared to control. Earthworms affect the chemical composition of soils and distribution of plant nutrients in them. Faecal material (casts) which consists of finely divided and thoroughly mixed organic and inorganic materials, is deposited on the soil surface, or in their own burrows or in other space below the soil surface resulting in the formation of organic matter enriched (humified) soil horizons.

Mortality of earthworm (dead earthworm tissue) also contribute to the soil, plant nutrients derived from metabolic and decay processes are recycled in the plant / soil ecosystem.

Casts of earthworm contains 75 to 80 per cent more organic carbon (high %) than undisturbed soil, representing an enrichment of surface soil. Parle (1963) found 6 to 7 times as much polysaccharides in casts as in undisturbed soil.

Nitrogen in cast is of plant tissue, is added in soil. Most of the nitrogen in earthworm tissue is incorporated into proteins. The quantities of nitrogen that may be cycled through earthworms appear to be extraordinarily large compared with the amounts generally recognized as being available for plant growth. Rapid recycling of \nitrogen through earthworms microbial tissue back to earthworms Nitrogen fixing micro-organisms like Rhizobium bacteria are common in the gut contents of earthworms. Their presence indicate that there is a true symbiotic relationship that promotes nitrogen fixation (French et al. 1976. Lee 1983).

Only small quantities of phosphorus (P) are excreted by earthworm in liquid wastes (Bahl 1947) but considerable quantities of the order of 5 to 10 times the available P in casts compared with that in surface soil.

pH of the earthworm casts is generally closer to neutrality (6.5 to 7.5) than that of soil from which they are derived.

Casts have higher concentrations of exchangeable Ca, Mg & K than the underlying soil. The increases in available cations are related to the higher content of plant tissue in casts than in soil. Concentration of readily available nutrients (Cations P,N) in casts provides a favourable environment for seed germination and plant growth. Earthworms possess a number of mechanisms for immobilization and excretion of heavy metals (Pb,Cd, Zn,Cu, Mn, Fe) (Bouche, 1983) and Iveland (1979). Shri Chande & Phatak (1951) reported that earthworm casting increased the availability of nutrients. Nijhawan & Kanwar (1952) observed that the application of earthworm compost increased available N,P & K and decreased the pH of soil as compared with the control. Swarup (1984). Jambhekar (1990) reported that application of vermicompost increase the available N, P & K of soil.

Earthworm increases the $N0_3$ production by stimulating bacterial activity and through decomposition of their own bodies. Researches carried out at Fruit Research Station, Ganeshkhind, Maharashtra revealed that organic carbon available NPK increased due to application of FYM and vermicompost. FYM and Vermicompost are at par with each other Anonymous (1992). Hapse et al (1993) reported that application of vermicompost @ 5 t ha^{-I} significantly increased total N available NPK, organic carbon and decrease pH over control.

Soil health is improved by the application of Vermicast and vermicompost as a super manure or biofertilizer (Lee, 1985). The pH of soil has decreased from slightly alkaline to neutral, available NPK were increased, decrease in salinity, increase in water holding capacity were observed due to application of vermiculture to soil (Gunjal and Nikam 1992).

ROLE OF HUMUS FORMERS

Eisenia foetida, Eudrilus euginea, are humus former and surface dweller in nature. Another humus former, *Perionyx excavatus* is of Indian origin. These Worms are less sensitive to change in temperature and moisture. Originally they feed on animal manure. They ingest in

90 % organic matter (half digested) and 10 % soil. They fed prey to predators like rates, moles very easily, as they live near the soil surface. Their life cycles is not more than 1 to 2 years, hence, their reproduction rate is very high. More than one generation is produced in one year.

VERMICOMPOST SHED (HEAP METHOD)

Commercial vermicomposting is a time bound process in view of the waste material storage, labour utilisation and timely market supply. Further the processes of decomposition by microorganisms as well as earthworms need some uniform conditions such as absence of sunlight, low temperature, and adequate moisture content. Earthworms do not tolerate U. V. rays and need protection from predators. Regular but cheap sheds are therefore, necessary for running a commercial vermicomposting unit. Under any circumstances, composting is an aerobic process by spreading waste in trenches or pits, the height of the heap being about 30 cm – 45 cm above ground level. With a view to use labour for turning and collecting Vermicompost at the end, the width of trench or pit is kept 1 m with a view to accommodate 2 such trenches or pits, and 0.6 m wide passages in between two pits the width of the shed 1m. The shed length can be from 18 – 30 m depending upon the space available. It is estimated that a heap of 1 m width, 15 cm deep and 0.6 m above ground level should accommodate in 18 m length about 4 tonnes of waste with about 60% moisture.

50 % black shedding net or low cost asbestos is an ideal cover for roofing. In case of heavy rains or leakages inside can be covered with white HDPE cloth. Overhead drip pipe with suitable nozzles can be provided for water supply. Such shed will help to keep a difference of minimum 5^0C from out side temperature. It's cost would vary according to material used and facilities provided.

A minimum of two sheds one for vermicomposting and other for earthworm working should be constructed later on, half digestion can be done outside and all pits be used of vermicomposting. A store for vermicompost storage inputs like bio-fertilizers, neem cake, etc. and empty bags, beans, etc. should be provided. Entire complex can be fenced by trees like cassurina neem, *Glyricida maculata*, acorus, and other plant species which would repel rodents insects and lower odour intensity. If overhead drip watering is not possible, 1.2 cm rubber holes can be provided for an unit of two to three sheds.

Essential inputs

(i) Earthworms: For a beginner, worms of species *E. foetida* and *E. euginae* are recommended. Since for a commercial unit a large number of earthworms will be required to begin with. It is necessary to multiply a small quantity say 6 Kg of worms for about a period of two months. This would give training and confidence and increase the worm number considerably. The worms at this stage may be fed cattle dung and crushed maize and dal in small quantities in ratio of 10 : 3. The vermiculturing on a small scale is a key necessity of the unit. Even when the unit is regularly running in full scale at least 2 Kg. (around 2000-2600) worms per pits must be kept under multiplication for use in emergency.

(ii) Cow dung : Both the species are dung dwelling and need addition of cow dung if waste is other than it. Small quantity of dung, stored for about a week or its slurry is needed for mixing / spreading. One can reduce the proportion of dung to be mixed as the worm number increases but a small quantity is always needed. Cattle dung and agricultural waste like grambran, kitchen waste, rice polish, sewedge, sludge, vegetable waste, wheat bran, etc. can be tried in the ratio of 10 : 3.

(iii) Compost culture : Old well rotten compost or garden soil as a starter is used helps early decomposition. If wastes like kitchen waste, which putrefy early, or wastes which takes time to decompose are used for vermicomposting, compost culture is required to avoid bad smell, volume reduction and early decomposition.

(iv) Neem and Karanj cake : Addition of karanj cake increases worm number and activity. Both cakes should be mixed @ 5 Kg. per tonne of waste, at the time of releasing earthworms. Besides correcting C : N ratio increasing number of worms, its repellent action is helpful.

(v) Soil or riversilt : Earthworm need small amount of soil for intake, this helps decomposition activity, besides escaping Nitrogen can be held by soil particles. 25 Kg. soil can be added / tonne of waste material.

(vi) Water : The percentage of moisture in the waste is to be maintained between 50 to 55, depending on its quality.

(vii) Gunny sacks as cover : The gunny sacks are needed to cover the heaps when earthworms are added. These cut off light, keep moisture level high at the surface which induce the worm to come to surface.

PROCESS OF VERMICOMPOSTING

(i) Partial decomposition of waste material

Waste separated from plastic, glass and metal pieces, should be used. From the nature of waste, its approximate C: N ratio can be judged. This will help in adding dung, cow urine, biofertilizers etc. The raw waste can be spread in long heaps in the shed. Every layer of about 0.2 m should be sprinkled with small quantity of superphosphate in the form rock phosphate starter compost, decomposing culture and drenched with water. If needed, azospirillum can be added. Height of the heap will be finally about 3.6 m. This should be sprayed with water at least once in a day. Slowly the heap will start heating within about 4 to 6 days.

(a) Mesophilic Stage: The soluble components of organic material will start break down with small increase in the temperature upto 35°C. The stage would last for 5 - 6 days. No turning should be given at this state.

(b) Thermophilic Stage: Soon the heat generating bacteria increase and the heap temperature will rise at least for a couple of days to 60°C or even above. The rise in temperature will depend upon moisture, aeration and types of wastes. Peak temperature may reach within about 10-15 days and then start lowering. The entire stage may take 20 to 25 days. It helps, to kill weed seeds and unwanted insects if any. pH (7.0 – 7.5) should be checked and correction accordingly should be made.

(c) Cooling phase: If there is no uniform decomposition, the temperature may rise once or twice again. Temperature may be recorded daily. Once the temperature reached around 30°C the material is fit for release of worms. If material is outside shed, it should be shifted to shed and spread in layer. Each layer should be sprinkled with dung / slurry, Karanj and Neem cake, slurry mixed with bacterial cultures, and soil. The heap should be raised to a height of 0.75 m at centre, above ground level. The heap should be observed for at least 2-3 days for change in temperature if any. If it does not rise much beyond 30^0C it is time to release worms.

Release of earthworms

If temperature and pH are normal worms can be released. The

heap should be covered with gunny sacks or cloths which should be made wet. Daily checking of the temperature (at the centre of heap) and study, if worms are still in the waste. Continue the observation for 3-4 days and if worms are found working, give cast on surface, then add another lot of about same number of worms. Thus for a new unit about 8,000 to 10,000 worms (8 -10 kg) for 18-20 m length are adequate.

For the next 25 to 30 days, no turning are to be given. The gunny sacks are to be kept wet but excess moisture must be avoided. Record of Datewise temperature must be kept. After about 15 days a layer of cast will be seen on surface. If heap is very compact and aeration is poor, the surface can be gently stirred with iron forks. Moisture should be adjusted.

Release earthworms at maturing phase of composting, proper micro-organisms including actinomycets, mycorrhiza, fungi and bacteria help decomposition and enrichment. Earthworm pass the waste through their bodies and give out the excreta as cast. The period for maturing the compost is about 30 days for partial decomposition and about 30 to 35 days for worm working. Depending on the number of worms, their type & technique used, the time may be little more. for commercial unit one should control the time for giving space to new batch of waste material. Mature vermicompost smells sweet as uprooted groundnut plants. It is powder mixed with cast, black in colour, non sticky.

Collection of vermicompost and worm separation

Once the vermicompost is ready in the heap, watering should be stopped. After two days the gunny cover should be removed. The worms will enter to the bottom of the bed. The vermicompost in trenches or pits should be piled in conical heaps with out much disturbing the worms in the bottom bed. The worms in the conical heap will collect at the bottom. After leaving the heap for a day the compost should be lifted for sieving on 1 cm mesh sieves. Every worker should be provided with plastic tray or bowl for collecting the worms finally a ball of worms will be found at the bottom, which should also be collected. All the worms so collected will be send in vermiculture section for feeding and multiplication.

Unsieved portion of compost, after separating the stones, bricks, etc. be kept separately for using as starter culture.

The vermicompost must have moisture to the extent of 60 %. The heap should be stored in shed for day or two, then if moisture content is correct, it should be bagged in 50 Kg. HDPE bags should be used, stitched and stored in shade. All earthworms, say, in gunny covers, near sieving or on the passage should be collected for reusing. Earthworms if allowed to be dried, in short time they will die. While in trays these must be in dung or wet vermicompost.

Vermicompost yields

Yield depends on the type of waste material and in C : N ratio from good material like cattle dung with 85% moisture. About 33% vermicompost at 55% moisture is considered good yield. From other wastes it may be between 25 to 30% of original waste material. Recovery of vermicompost will be of 33%.

Quality of vermicompost

Black powder with pleasant smell, low in density, and giving almost clear liquid when mixed with water is taste of good vermicompost.

NUTRIENT STATUS OF VERMICOMPOST

Sharma & Sharma (1983) reported earthworm casting contains as much as 5 times more nitrate nitrogen, 14 times more calcium, 3 times more magnesium, 11 times more potassium than that of 15 cm top soil. Krishnamorthy and Vajranabhaih (1984) noticed *Pheretima elongate* released 83 nano-grams of cytokinin per day per se and 419 nano-gram auxin (growth regulators) per day per se, which are present in intenstine of worms.

Nutrient status of vermicompost increasing different organic waste as found source may be as follows - organic carbon 9.15-17.98%, total nitrogen 0.5-1.5%, calcium and magnesium 22.7-70.0 mg/100 gm, copper 2.0-9.5 ppm, iron 2-9.3 ppm, zinc 5.7-11.5 ppm and available sulphur 128-548 ppm (Kale 1998).

Shinde et al. (1992) reported that vermicompost contains more carbon and phosphorus than FYM less potassium and micronutrient than FYM and had comparable contents of nitrogen. It has wide C : N ratio as compared to FYM.

The average nutrient component of vermicompost and FYM is generally as follows :

Nutrients	Vermicompost	FYM
Nitrogen (%)	2.5-3.0	0.5-0.93
Phosphorus (%)	1.5-2.0	1.0
Potash (%)	1.5-2.0	1.31
pH	7.0-7.5	7.2-7.9

REFERENCES

Gupta, P.J.C. 2003. A Hand Book of Soil, Fertiliser and Manure, Agrobios India. Jodhpur. pp. 366.

Krishnamoorthy, R.V. and Vajranabhaiah. 1984. An assessment of plant growth promoters level in worm casts. A paper presented to National Seminar on Organic Utilisation and Verraicomposting, School of Life Sci., Sambalpur Univ. Grissa, Dec. 5-8, 1984. pp. 8.

Kale, R.D. 1998. Earthworm Cindrella of organic farming. Prison Books; Pvt. Ltd., pp.70.

Sharma & Sharma. 1983. Earthworms for soil health and population control. J. Sci. Ind. Res. 42: 575-583.

Shinde, I.P.H., Naik, R.L, Naszirkar, R.B., Kadaro, S.C. and Khaire, V.M. 1992. Evaluation of vermicompost. Proc. of National Seminar on Organic Farming held at College of Agriculture, Puns from April 18-19. 1992. pp. 54-54.

❑❑❑

7

Biodynamic Farming : A Holistic Approach in Agriculture

MOHIMA PRASAD BEHERA AND PRAVAT KUMAR ROUL

India's values and traditional healthcare system is based on Ayurveda is as old as Vedic era dating back to Rig Veda (4500-2500 BC). In Atharva Veda (2500-2000 B.C) clear description of various medicine plant uses have been elaborated. In Samhita period , the science of medicine systematically organized with a clear concept and theories have been picturized in the Charak Samhita -2000B.C and Sushruta Samhita -1000-800 B.C. The Veda and Upanishad are and said to be holy scriptures followed by all scholars' of all religion throughout the world. Therefore the knowledge of spiritual values, traditions and holistic approaches towards ancient medicinal plants cultivation were known to ancient Aryans going back to the period of 4500-2500 B.C.

Let us discuss how spiritual values and holistic approaches can be related to similarities between growth of a food grain and growing of a man from animal instinct to divinity.

For growing of a grain we require a suitable field with proper soil condition like good fertility. When we want to reach divinity we need a suitable field for our proper growth and development. The field is nothing but heart filled with purity, softness and devoid any bad intention or stupidity. This helps a man growing from animal instincts to a divine being. Therefore when our heart is hard, full of stupidity and jealousness, it can never reach the desire gol . It is just like a deserted , barren land full of hard pans unsuitable for proper growth and development of a grain. This is how the famous dacoit Ratnakar turned to divinty and became a Great Rishi Valmiki, the writer of holy epic Ramayan. The divinity of the seed is expressed in better yield.

It is generally observed that if any plant is loved , taken care for and closely observed minutely its divinity is expressed in more branching , flowering and fruit setting. The hearts divinity expression is transformed to cosmic energy through mind which make the environment suitable for growth and developement. That is why the Rishi's and ancient people used to chant mantras, use tantras while sowing plant according to different position of sun or moon which creates energy field at the ground level for its better utilization and transformation when agricultural concept did not exist.

HISTORY OF BIODYNAMIC AGRICULTURE

Dr Rudlof Steiner founder of anthroposophy in the Germany (now in Poland) used to research and investigate forces relating life and growth.He developed a unified approach to agriculture relating ecology of the earth organisms to that of the entire cosmic in 1924 on the request of farmers who noticed degraded soil condition and deterioration in health and quality of crops and livestocks due to use of chemical fertilisers, pesticides, synthetic growth regulators, etc. The principles of biodynamic farming is based on developing life giving humus out of dead organic material being decomposed continuously. This biologically active humus is the basis of all living soil. Humus is the slowly available nutrients supplied to plants for their proper growth and development

Biodynamic agriculture is a method farming aiming to treat the farm as a living system which intereacts with environment to build healthy and living soil and to produce food that nourishes and vitalize helping development of mankind.Simply it is making life giving humus composed out of dead material.

Steiner's gift to the world is knowledge from the spiritual world that enables people to recreat and work with natural forces in the field of education , medicine and agriculture. Steiner taught that "Matter is never without spirit to the vital cosmic energy and spirit is never without matter."

United States Biodynamic Farming and Gardening Association was found in 1938 as a New York State Corporation. In Australia first biodynamic preparation was made by Ernest O Genonion in Melbourne in 1927. Now Biodynamic Research Institute (BDRI) is situated in Melbourne.It is practiced in more than 50 countries worldwide.

THE ORGANIC CONCEPT

Steiner advocated discontinuing the use of chemical fertilizers, pesticides, synthetic growth regulators that pollute the environment leading to degradation of soil fertility, reduction of population of soil living organisms. These materials also lack inherent life ingredients and forces. The crop that we grow and foods quality we consume is totally depended on the cosmic energy that comes to a living soil supporting our daily life.

Working with cosmic formative forces is the electromagnetic radiation mostly in the microwave range is believed to be highly redshifted effect of the explosion billions year ago from which according to BIG BANG model, the universe was created. Rodert Wilson Arno Penzias discovered this in 1964 accidentaly. The cosmic rays originate from sun and outside solar system i.e. Galaxy's disc and extragalactic. Earth have very high energy interception. These rays are of high quality of speeds of about 87% of that of light.

Steiner opined that plants grew not only through the soil fertility but also support from the cosmas the rhythms of the sun, moon, planets and the wider constellation of the Zodiac. The cosmic support spreads into the earth through the living nature of soil which is vital for healthy soil yielding quality food.

Steiner's principle is based on the hypotheses that as earth becomes older the soil naturally becomes weaker and is able to recept at lasser rate at which cosmic rhythms received in the past period. He empathized on a renewal of agriculture to revitalize the soil force, so that the cosmic energy could stream again strongly into the soil as before. He introduced the practice of natural preparation based on cow manure, silica and various herbal plants that open up the soil to those influence again.

He strongly laid down intimate connection between all living things and the cosmos and the importance of the workings of the planets and heavenly bodies.

BIODYNAMIC FARMERS

The concept of farming revolves around integration of crops and livestocks recycling of nutrients, maintenance of soil and the health and well being of crops and livestock. The farmers too forms a part of the farming. In this system, cover crops, green manures and crop rotations are used extensively.

BIODYNAMICAL PREPARATIONS

Steiner recommended nine different preparations to aid fertilization which is the main component of the biodynamical agriculture. He opined that these preparations transformed supernatural terrestrial and cosmic forces in to the soil. The preparations are numbered BD 500 to BD 508. The preparations 500 and 501 are used for preparation of the field and rest are used for making compost. The following observations are generally carried out.

- Spraying of same preparation stimulating plant growth regulators including Cytokinins.
- Improvement in nutrient content of compost.

Field preparation (BD 500-501)

The BD 500 and 501 preparations advances humus formations.

- BD-500 (cow horn dung). It is prepared by burying cow horns filled with cow dung in the soil for six months. During autumn and winter, cow-dung has the amazing ability to increase the population of earthworms, the natural soil nourisher. According to Steiner the cow horns have the ability to hold cosmic ray and helps on cow's proper digestion resulting in high quality of dung. When the horns are removed from the cow, they still retain the cosmic ray and have ability to absorb life energies that are in the soil in the winter during the decomposition of dung in horn.

Major ingredients

- **The Dung :** It must come from healthy cattle fed on pastures grown on biodynamic method.
- **Soil Quality:** The horn should be buried in vegetable garden when sustainable management is practiced continuously with BD compost, green manuring and crop rotations.
- **Cow horns:** It should be locally available, small to medium in size not larger than 5cm in diameter generally preferred.
- Pit:- It should be within fertile top soil layer. It should be buried 75-150cm depth down the soil. The size of pit should be 2x4x0.4m.

Method of application

Preparation 500 should be applied at least twice in a year once in springs and once in autumn. It is applied first on autumn (Aug-Sep) in northern hemisphere and March- May in southern hemisphere. Second application is done in Springs (May- June) and Sep-Oct consecutively. Normally BD 500 is applied in a descending period of moon rhythms. (14days)

Quantity

The rate of application is 25g preparation in 13 liter of water per acre and 65g in 33liter per hectare. A 1000 acre farm may require 12.5g in liter of water per acre.Spraying of BD 500 should be made up to a solution under water application. Before applying it is to be stirred first one hour before BD501 spray i.e. 1g mixed in 13 liter of water for one acre spray.

Preparation of BD 501

It is made from cow horn. The horns are filled with finely powered quartz and are buried at the opposite time of the year from preparation 500. The BD501 remains in earth between springs and autumn the "light" time of the year between March and Sep-Oct in northern hemisphere including India and Sep-Oct and March -April in the southern hemisphere.The action of BD501 is the light force i.e. released from the crushed quartz crystals. The extra light intercepted increases photosynthesis and encourage development of fruit and seed. Initialy BD500 is used for one year and thereafter BD501 is used next year since its main role is in light process and photosynthesis.

BD501 is made by taking a well formed clear quartz crystal (Silicon dioxide) and crushing it in a metal structure or with a hammer or in motor and pestle made of iron or steel. It is further put between two sheets of plate of glass to be further grinded and till it feels smooth by finger and thumb. The powder is made into a slurry with water and placed in a cow horn. The horn is left to stand overnight to draw out excess water form the top and horns are buried in earth on a day when the moon is ascending. The site of bury should be marked with proper label. Otherwise retrieving will not be easy.

Time of application

Spraying of 1g of BD501 in 13 liters of warm rain water is required for one acre done in early morning at sunrise in Springs.

Uses

Generally BD501 is used for fruit crops such as apples, peaches, plums and citrus, when the fruit attain the size of a shelled walnut. Each month one spray can be done.In case of cereal grains it is never used at the half way through plants normal growth as it brings premature flowering. In case wheat and other annual grain crops first spray is done at four leaf stage and second after flowering and second one before harvesting i.e. in matured grains when the moon is in Leo which is warmth and seed formation period.BD501 is important for strengthening seed quality according to Dr. Rudlof Steiner. The grains become hard and suitable for confectionary purpose. The sugar content grapes increased with use of BD501 from an average of 25 (Brix level) to 64 after spray in New Zealand by grape grower Gisborne in 1994.Since BD501 is a very powerful tool care should be always taken for its use by experts only.

Compost preparation

A well decomposed compost incorporation is important for maintaining and staining humus content, micro-organism and earthworm activity. It is the major cheap source of nutrient supply and enriching nutrient pool and conserving soil and moisture. The major two steps involves one pile heat up by which there is active and quick breakdown of protein at a temperature between 60^0to70^0which covers for a period of six to seven weeks. In the second stage a slower, stable humus formation occurs and worm activity increases.

Biodynamic compost preparation : (BD502)

In biodynamic compost preparation herbs used for medicinal purpose are frequently used BD 502 is prepared with yellow blossom (*Achillea millefolium*) which are stuffed into urinary bladder from Red Deer (Cervus ele phus) placed in the sun during summer, buried in earth in a bottomless box or earthen pit during winter and retrived in the spring. It doesn't form humus however, the flower remain intact and stored in a jar and the bladder gets rotted.

This preparation brings light forces to the soil in the form of potassium and sulphur. The yarrow plant grown on waste lands have a connection with sulphur and potassium. Its white composite flowers are connected by light forces with Venus. Each flower is like a chalice receipting beneficence of the cosmos.

BD503

It is made from the flowers of German Chamomile (*Matricaria recutita*) which are stuffed into small intestine from cattle burred in humus rich earth in autumn and retrieved in the springs.

BD503 promotes in easy breakdown of protein molecules in to compost humic plant nutrients and prevent breaking down to ammonia which get lost. It is main source of calcium.

Preparation

The fran cow's intestine is cut in to lengths of 25-40cm. after taking out the partially digested contents of the same by running them between forefinger and thumb. Through the open end of the intestine, a small amount of dried flower is put into intestine and tied in a strings and buried in autumn and lifted in the springs before earthworms remove it. All the other preparation can be dug up in spring except the nettle which is buried for 12 months.

BD504

Stinging nettle (*Urtica dioica*) plansin full flower are buried together underground surrounded by in all sides by a peat for a whole year. According to the Rudolf Steiner nettle is a plant with unique healing qualities for humans and plants. When a sketch of a nettle plant is made and line is connected to a nodal points of axils of each leaf, a double spiritual stuff of Mercury is found i.e. the medical sign of medicos. The forces of Mars influencing nettle flower brings iron, magnesium and sulphur into soil. The nettle preparation helps to break up iron pan soil, releasing iron.

Preparation

The harvested nettle before flowering in late spring or early summer is dried and kept as such till autumn. The dried plant is stuffed into a round unglazed earthenware tile in field at a diameter of 10-15cm. The plant is firmed into the plant in a tile with the handle of a broom and buried in good soil in autumn for 12month. The covered top soil is marketed well with a label. After a year the nettle shrunks to a 2.5 cm of black compost at the bottom of tile. It is colloidal in nature. It is sieved to break up the stalks before strong in jars.

BD 505

It is prepared with oak bark (*Quercus robur*) which is chopped into small pieces and placed inside the skull of a domesticated animal, such as cow, sheep, goat surrounded by peat and buried in the place where lots of rain water runs past.

According to Rudolf Steiner soil should have Ca as an important living form to make the plant healthy and free from disease. Here the calcium is not applied as a minor. Bark Oak contains 2.5-5% Ca. According to Steiner the skull is used as it is formed first with moon-growth process. He opines that moon- growth process is seed germination phase which is enhanced at the time of full-moon.

Preparation

It is made in autumn. The oak bark is griended in to fine powder then packed tidely in to a domestic animals skull. It is buried in earth when rain water is supplied with a down pipe from a barrel surrounded by plenty of rotting smelly vegetation (leaves, grasses) and vegetable foliases. This skull is retrieved in spring and split to take out the preparation to put in glass jar. It smells because of anaerobic decomposition. It is turned everyday until aerobic activity takes over and smells quite sweet. The time taken for completion a is fortnight.

BD-506

Dandelion flower (*Taraxacum officinale*) is stuffed in to the peritoneum of cattle and buried in the earth during winter and retrived in spring. The flower of Jupiter symbol having a crystalline silica cosmic form sphere. The flowers are picked in the morning. The flowers are dried and stored. The fat is cutoff from the mesentery that holds the digestive organ of cow. The flower is piled into it and skin is folded to make it a tight "parcel" which is tied up with string and buried in the soil in descending phase of moon in autumn. The preparation is retrieved as in case of BD500. The mesentery is cut open. The bulk preparation is removed and stored in a glass jar for few weeks for aerobic activity. The preparation gives the soil the ability to attract as much silicic acid from atmosphere and form the cosmos needed by plants. Silicon is thought to be essential for phosphate fertility of the soil. In the form of silicate, it displaces phosphate from unavailable compounds so that they became available to plant. Inside the plant it counteract the toxicity of iron and cupper. It protect the plant from attack of fungi and aphids. In thick stands of rice, it prevent lodging.

BD 507

It is prepared with valerian flowers (*Valeriana officianalis*) extracted in to water. Its concentrate phosphorous which is a vital constituent of plants as it is involved in attracting the light used in photosynthesis BD 507 stimulate the phosphate process and mobilises the phosphate- activating bacteria in to bring the Saturn influence on worms.

The flowers are harvested in summer in an ascending period of moon. The flowers are grinded in pestle and mortar and kept in jar with four times their volume of distilled water (20% solution). The jar is left in a window sill for seven days and filtered and kept air tight and in a dark cup board for used.

Method of Application

A 5% solution is made and 10ml is stirred into 13.5 liters of rain water for 10 minutes and applied @13.5 liters per acre. The valerian is not put inside the compost like other preparation but spread over the top. Earthworm activity raises soil temperature by 1-2^{0}C protecting sensitive plants from frost. Valerian solution are sprayed over grains to improve germination.

BD-508

It is prepared with horsetail plant (*Equisetum arvense*) grown in swampy wastes in Europe and Austria. it has a very high silica content.

Preparation

The plant is harvested in spring, dried and added with rain water @ 100g per 2 liters of water. It is boiled and let in simmer in covered pan for 20 minutes. it is left as such for two days.18 litres of water is added and stirred as BD 500 for 10 minutes.

Uses

The BD508 is spayed in the ground to protect it just before stress timese i.e. just before full moon or perigee when the moon is closest to the earth. it is sprayed regularly in springs.

APPLICATION OF BD COMPOST PREPARATIONS TO COMPOST HEAP, LIQUID MANURE AND COWPAT PIT

Any compost heap and liquid manure and cowpat pit preparation is incomplete if not finally applied with any BD compost. Since it helps in decomposition process, increases the biological activity, and life forces. Any one set of preparation is applied @ 1gm of BD 502-506 and 10 ml of BD 507 per 5 cubic meter(5tonnes) of compost heap per 200 liters of liquid manure. These three sets of BD preparation is required for 60 kg of cowpat pit.

In heaps, through deep hole spaced out, solid BD preparation are pushed inside and closed well. In two separate holes on top of heap, one quarter of stirred valerian is poured and rest liquid is spread around the heap.

In case of liquid manure , each BD preparation is wrapped in old compost or dried decayed glass and float on top of liquid manure (fish manure)BD 507 is sprinkled over the top after being stirred for 5-10 minutes.

The BD preparation are pushed in the cowpat pit manure (one barrowload of 60 kg of firm cow manure +200g crushed egg shell + 500 gm of basalt meal or dust mixed throught out and put at a depth of 250mm pit in a vegetable garden area for three months) and valerian is sprinkled over the top.(10 ml of 5% valerian solution + 500ml of water).

Astronomical planetary calendar

The astronomical influences on soil and plant development as ascending or descending period of moon's effect on solar light interception or life forces on the planet is most appropriate for scheduling plantings,dates, cultivation or harvesting times of various kinds of crops is described in the biodynamic agriculture Calendars for every year.

Significances of biodynamic farming

The biodynamic compost(BD) preparation play a significant role in unified approach to agriculture. It is a complete substitute for pure organic agriculture beings biological and ecological in nature. it's main composition are of medicinal herbs, cowpat manure and silica of resulting in a long process of fermentation in order to enrich the growth stimulating substance. The preparations for 500-508 are applied to manure and compost piles in very small quantities. It avoids uses of chemical fertilizer, chemical herbicides , pesticide and growth substance polluting the environment, degrading soil health and increasing human hazards.

Biodynamic is a way of living , workings and completely related to nature and the vocation of agriculture based on good commonsense practices and awaring the conscious of each practitioners for self-sufficiency in energy, nutrient applications to plant and animal. it is a total ecological agriculture practices which is in tune with nature round the clock increasing the lifespan biological clock of individual plant and animal and the farmer. Its contribution to the understanding of the astronomy, soils ,climate, plant and animal life ,human ecology, and economy of a particular locality can not be ruled out.

ORISSA - A STATE OF HOLISTIC APPROACH IN AGRICULTURE

A common odiya farmer starts his farming operation like sowing of seeds from the auspicious day of Akshya Tritiya - the day of Baisakh Masha Sukla Pakhya (ascending period of moon). On that day he takes the seeds in a decorated box with incense sticks, vermillin coconut and fruits for worship by chanting by mantras. He wears new clothes and observe the day with delicious dishes and in the evening Krishi Bhagabat (a holy agricultural poem) is recited in groups. This occasion falls in the April - May month indicating the approach of monsoon and he hopes that seeds will germinate fully on receipt of a good monsoon and attain divinity with good harvest.

Again in the holy Raja Sankranti's 4th day observations in Jaisthya Mash (June), all field operations like hoeing, weeding, intercultural operations, sowing, etc. are stopped and implements are cleaned, decorated and worshipped by chanting mantras (syallbles) hoping safe operation of the same and a good return from soil. During this period there is greater movement of Sun and a good reception of solar energy at ground level.

In Gamha Purnima in Sravana Mash (full moon day - August month), the bullocks are decorated with new coloured clothes, bells, vermilin and worshipped for their well being and physical fitness throughout the year so that all field operations can be carried out smoothly.

In Margashira Mash's every second to fourth Thursday, (November - December month), Goddess Laxmi is worshipped hoping a bumper harvest of the crop which concides with harvesting festival of Bihu in Assam and Pongal in Kerala. He hopes a good economic return from his crop and observes every thursday in dancing and cultural programmes on the occasion.

Hence it can be concluded that all the farmers practise holistic approach in agriculture as a way of life rather than an earning for livelihood.

REFERENCES

Britannica Ready Reference Encylopedia vol. 3, 2006. Publish by Encyclopedia Britannica (India) Pvt. Ltd., New Delhi – 2000 pp 58-59.

Proctor, P. and Cole, Gillian. 2000. Grasp the Nettle (Making biodynamic farming and gardening work) published by Random House, Glenfield, Auckland, New Zealand.

Appendices

APPENDIX-1

Timetable for Making and Applying Preparations in Different Parts of the World

These calendars provide a rough guide for beginners. For successful bier dynamic management, farmers need to develop a 'feel' for Nvhen is the right time to apply the preparations for their particular farm and climate.

In hot countries the fermentation time for all preparations will be shorter than in temperate climates: in warmer climates it will take 4-5 months; in cooler climates, 6-7 months.

TABLE Al

A Calendar for Making and Using Preparation 500

Action	NZ & temperate Australia	India, tropical Australia Asia	Europe, northern US & Canada	California & Mediter anean
Bury	April	Oct/Nov	Sept/Oct	Nov
Lift	late Sept	Feb/Mar	Nlar/May	Mar/Apr
Apply[1] in spring	Sept/Nov	Feb	late May/ early Jun[2]	April
in autumn	Mar/May	*a*-t/Nov[3]	Aug[4]	Oct; Nov

1 Apply during descending moon.

2. Apply,when soil temperature rises.

3 Apply after the last monsoon rains.

4 Apply before soil temperature falls.

TABLE A2

A Calendar for making and Using Preparations 501-508

Prep	Year	Spring	Late spring/ summer	Autumn
501		bury horn		dig up horn
502	1	make & hang	harvest & wilt flowers	take down & bury
	2	dig up		
503[1]	1		harvest & dry flowers	make & bury
	2	dig up		
504	1		harvest foliage	bury
	2			dig up
505	1			gather bark, make & put in water
	2	retrieve		
506[2]	1		harvest & dry flowers	make & bury
	2	dig up		
507			harvest flowers & make	
508		harvest & dry		

1. In most countries the flowers can be harvested and the preparation made in the late spring. Where flowers are unavailabie by late spring, gather them the previous vear, drv and store.
2. You could also try making it the previous ious spring and hanging it in a tree for 6 months, as for Preparation 502.

Note: compost, liquid manures and cowpat pit can be made, and the compost preparations added to them, at anv time of the year, although decomposition is quickest during the ,%-armer months of the year.

Steiner suggests hanging up the Preparati-on, 503 and 506

APPENDIX-2

Biodynamic Associations : World Wide

Biodynamic associations throughout the world provide some or all of the following services for their members:

- information and advice
- sale of preparations
- directories of suppliers for various resources
- farm inspection and certification to Demeter standards
- educational seminars, discussion groups, courses in biodynamic methods

New Zealand

The Bio Dynamic Farming and Gardening Association in New Zealand (Inc.) PO Box 39045, Wellington Mail Centre, New Zealand. Ph +64 4 ,589 5366

e-mail biodynamics@clear.net. nz

Australia

The Biodynamic Farming and Gardening Association in Australia (Inc.)

PO Box 54, Bellingen, NSW 2454, Australia.

Ph +61 2 6655 8551

e-mail poss@rnidcoast.com.au

USA

Biodynamic Farming and Gardening Association (Inc.) Building 1002B, Thoreau Centre, The Presidio, PO Box 29135, San Francisco, CA 94129 0135.

Ph +1 415 561 7797

e-mail biodvnamicQaoi.com

Canada

Bio-Dvnamic Agricultural Society of British Columbia, Skimikin Road, CDN-Chase, BC VOE IMO,

Ph +1 604 679 S052, fax +1 604 679 2932

Hawaii

Biodynamics Hawaii. 2514 Alaula Way, Honolulu,

HI 96822, Hawaii, USA.

Ph +1 808 988 4555

e-mail pdwyer@hawai.i.rr.com

United Kingdom

Biodynamic Agricultural Association, Painswick Inn, Gloucester Street, Stroud, Gloucestershire GL5 1QG, UK.

Ph +44 145 375 9501

e-mail bdaa@biodynamic.freesene.co.uk

India

Bio Dynamic Association of India, No 78, 1st Floor, 11th Cross, Indiranagar 1st Stage, Bangalore, Karnataka, India.

Ph +91 80 5272 185

e-mail imoind@blr. N,snl. net. in

Germany

Forschungsring ffir biol.-dyn. Wirtschaftsweise, Brandschneise 2, D-64295 Darmstadt

Ph +49 (06155) 84690, fax +49 (06155) 846 911

e-mail d emeterbd @-online. de

Ireland

Bio-Dynamic Agricultural Association in Ireland, Ballinroan House, IRL-Kiltegan, Co. WickloNv, fax +353 508 81006

South Africa

Bio-Dynamic Agricultural Association of Southern Africa, PO Box 115, Paulshof, South Africa 2056, ph +27 11 803 1688

There are also associations in Quebec and Ontario in Canada, and in Austria. Brazil, Denmark, Finland, France, Israel, Italy Croatia, Luxembourg, Netherlands, Norway, Sweden, Switzerland, Slovenia and Spain.

APPENDIX–3

SIGN FALL

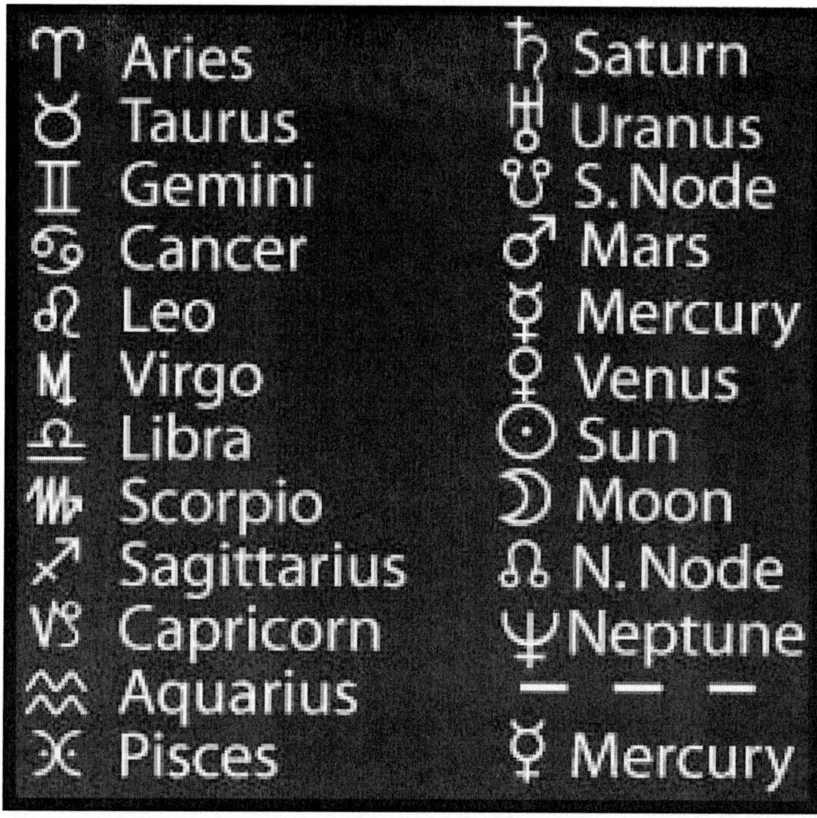

APPENDIX–4
Biodynamic Agricultural Calendar (Hindi-2001)

BHADRAPAD SHUKLA-PAKSH — 20 AUGUST – 2 SEPTEMBER 2001

SUN ☉ SINH  →

MOON OPP. SATURN ☽☍♄ 26.8 22.30

☾	1			🕑		☽ ☊☋ AgPg
∨	20 MON	SINH	♌		FRUIT	
∨	21 TUE	KANYA	♍	2.30	ROOT	
∨	22 WED	KANYA	♍		ROOT	
∨	23 THU	TULA	♎	6:30	FLOWER	
∨	24 FRI	VRISHICK	♏	14:30	LEAF	
∨	25 SAT	VRISHICK	♏		LEAF	
∨	26 SUN	VRISHICK	♏		LEAF	
∨	27 MON	VRISHICK	♏		LEAF	
∨	28 TUE	DHANU	♐	1.30	FRUIT	☋ 14.30
4.30	29 WED	DHANU	♐		FRUIT	
∧	30 THU	MAKAR	♑	20.30	ROOT	
∧	31 FRI	MAKAR	♑		ROOT	
∧	1 SAT	KUMBHA	♒	21.30	FLOWER	
∧	2 SUN	KUMBHA	♒		FLOWER	Ag 4.30

8

Organic Farming and Biotechnology

PARUL SHARMA AND P.C. TRIVEDI

"Organic agriculture is a production system that sustains the health of soils, ecosystems and people. It relies on ecological processes, biodiversity and cycles adapted to local conditions, rather than the use of inputs with adverse effects. Organic agriculture combines tradition, innovation and science to benefit the shared environment and promote fair relationships and a good quality of life for all involved."

–International Federation of Organic Agriculture Movements

A type of farming that restricts the use of chemical fertilizers and chemical pesticides and relies more on crop rotation and green manure is known as organic farming. In Europe, organic farming is known as ecological agriculture. Organic farming helps to sustain the mineral resources of the soil. There has been an increased demand for organic products such as cotton clothes including baby clothes, food and so on. Today, many countries are promoting organic farming based on the standards and norms set by the International Federation of Organic Agriculture Movements, an international organization that was established in the early 1970s. The concept of organic farming originated out of the organic movement that existed in the earlier years of 1930 and 1940. The organic movement was actually an anti-synthetic fertilizer movement. Synthetic fertilizers were used widely in those days for agricultural purposes whereas organic farming prohibits the application of synthetic fertilizers and pesticides. Organic farming is a very old concept and a developed practice in India. In fact, it was implemented in India much before the organic movement began. Evidences of this fact have been found in the ancient texts of India. In India, many of the villages still practice organic farming methods.

COMPONENTS OF ORGANIC FARMING

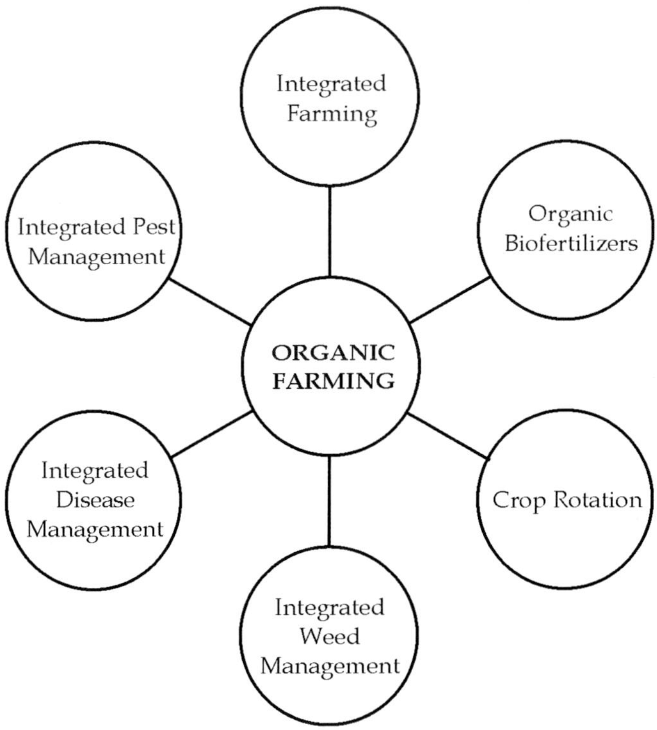

IS ORGANIC FARMING "ECO-FRIENDLY"

Energy

Currently, we use around 10 calories of fossil energy to produce one calorie of food energy. In a fuel-scarce future, which experts think could arrive as early as 2012; such numbers simply won't stack up.

On average, organically grown crops use 25% less energy than their chemical cousins. When these savings are combined with stringent energy conservation and local distribution and consumption (such as organic box schemes), energy-use dwindles to a fraction of that needed for an intensive, centralized food system.

Greenhouse gas emissions and climate change

The production of ammonium nitrate fertiliser, which is indispensable to conventional farming, produces vast quantities of nitrous oxide - a greenhouse gas with a global warming potential some 320 times greater than that of CO_2. The techniques used in organic agriculture to enhance soil fertility in turn encourage crops to develop deeper roots, which increase the amount of organic matter in the soil, locking up carbon underground and keeping it out of the atmosphere.

Water use

Agriculture is officially the thirstiest industry on the planet, consuming a staggering 72 per cent of all global freshwater. Organic agriculture is different. Due to its emphasis on healthy soil structure, organic farming avoids many of the problems associated with compaction, erosion, salinisation and soil degradation, which are prevalent in intensive systems.

Localization

Food transport accounted for more than 30 billion vehicle kilometres, 25 per cent of all HGV journeys and 19 million tonnes of carbon dioxide emissions in 2002 alone. The organic movement was born out of a commitment to provide local food for local people, and so it is logical that organic marketing encourages localisation through veg boxes, farm shops and stalls. As we enter an age of unprecedented food insecurity, it is essential that our consumption reflects not only what is desirable, but also what is ultimately sustainable.

Pesticides

A spiralling dependence on pesticides throughout recent decades has resulted in a catalogue of repercussions, including pest resistance, disease susceptibility, loss of natural biological controls and reduced nutrient-cycling.

Organic farmers, on the other hand, believe that a healthy plant grown in a healthy soil will ultimately be more resistant to pest damage. Organic systems encourage a variety of natural methods to enhance soil and plant health, in turn reducing incidences of pests, weeds and disease.

Eco-System impact

Organic farms actively encourage biodiversity in order to maintain soil fertility and aid natural pest control. Mixed farming systems ensure that a diversity of food and nesting sites are available throughout the year, compared with conventional farms where autumn sow crops leave little winter vegetation available.

Biodiversity is enhanced at every level of the food chain under organic management practices, from soil micro-biota right through to farmland birds and the largest mammals.

Food biodiversity

Seeds are not simply a source of food; they are living testimony to more than 10,000 years of agricultural domestication. Tragically, 75% of the genetic diversity of agricultural crops has been lost over the past 100 years. Modern intensive agriculture depends on relatively few crops- only about 150 species are cultivated on any significant scale worldwide.

Seed-saving and the development of local varieties is a key component of organic farming, giving crops the potential to evolve in response to what could be rapidly changing climatic conditions. This will help agriculture keeps pace with climate change in the field, rather than in the laboratory.

NEED FOR ORGANIC INPUTS

Several reasons have been emphasized for the need of organic agriculture like limited land holdings, poor socio-economic conditions of farmers, rise in input cost etc., the broadest view shows two major reasons. They are mainly, population and environment. They have emphasized the ultimate need for Eco friendly technologies. The current methods of agronomy are damaging to the natural resources and the environment. This type of agronomy also requires the farmers to invest heavily into inputs to improve yields and productivity. In addition to this many research studies have proved the presence of pesticide residues in vegetables.

- Recently Columbia reported for problems in reproductive processes as a consequence of consumption of the chemical methyl bromide through vegetables.

- About one million people suffer from pesticides poisoning.
- 20000 die every year due to the toxic effect of the chemicals used in agriculture.
- The crop loss due to pest is still 15000 crores.

In order to avoid the deleterious effects of synthetic chemical fertilizers and pesticides, organic agriculture is needed as an alternative to provide ecologically safe methods of farming.

The way the organic inputs are processed

Organic inputs are processed by a complex web of soil organisms; however biological management is based on the assumption that the outcomes are relatively consistent. These outcomes are multiple and include the generation of inorganic nutrients such as ammonium, nitrate, phosphate and sulphate from organic compounds of these elements.

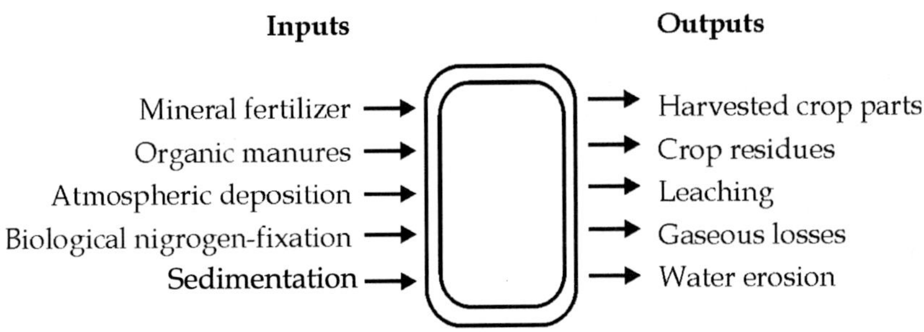

Much attention has been given to the use of organic mulches to supply nutrients such as Nitrogen and to the possibilities of utilizing resource quality control as a tool for managing nutrient availability. Resource quality refers to the regulatory effect of the chemical composition of an organic resource on its rate on pattern of decomposition and nutrient release.

PRINCIPLES OF ORGANIC FARMING

There are several compelling principles that characterize organic farming. Most of the farming operations followed will reflect all of these to a greater or a lesser degree.

- **Biodiversity:** Good organic farming reflects the biodiversity of nature through practices like intercropping, companion cropping, establishment of beneficial habitats, crop rotation etc., Farms with a diverse mix of crops have a better chance of supporting beneficial microorganisms that assist in pollination and pest management.

- **Sustainability:** Enterprise diversification helps in achieving sustainability in production.
- **Natural plant nutrition:** Crop rotation, green manure, farm manure, compost, natural fertilizers etc., should find applications.
- **Natural pest management:** On organic farms massive pest breaks are relatively rare and short lived due to the presence of natural predators, parasites and disease agents that quickly knock the pest numbers back down to a moderate level.
- **Integrity:** Integrity refers to the system in place and action undertaken to assure that consumers of organic products get what they pay for. The organic food items must be protected against contamination with that produced inorganic methods.

ADVANTAGES OF ORGANIC FARMING

- Natural resources will remain intact: The land and water resources which are on the verge of deterioration will be conserved.
- Pollution hazards: Land and water pollution is reduced, as pesticides do not find applications.
- By products of livestock and crops will be used wisely and appropriately: Animal dung can be used in the production of biogas and solid wastes can be utilized for organic manure to fulfill the nutrient needs of plants. The crop residues can be converted to biomass and can be utilized for plants.
- Provides employment opportunities: Ample labor is required for organic farms.
- Helps to generate original ecosystem: Original eco system refers to production, consumption and decomposition. No interference of chemicals with the natural harmony. Ecological balance restored.
- Cost of production is lower than modern farming: Agro wastes and vegetable scraps are used. Investment on chemical fertilizers is more a burden rather than a profit to the farmers.

DRAWBACKS OF ORGANIC FARMING

There are not many drawbacks in this farming practice. But they need to be considered to have a good yield. Following are some of the drawbacks of this method of agronomy:

- **Not nutrient specific:** Most organic manures are not nutrient specific.
- **Labour:** Ample labour is required; otherwise the land goes a waste.
- **Timely vigilance:** Regular vigil is required. As in the case of chemical farming, the pesticides once sprayed last for certain duration. Since Organic agronomy requires the complete discard of pesticides, a regular inspection of land is necessary.

Despite the drawbacks of this practice of agronomy, it can be regarded as a sustainable farming method.

METHODS OF ORGANIC FARMING

(a) Organic pest control

Organic pest control refers to the control of pests using organic methods. It causes less damage to soil and pests whereas chemical farming causes severe damage to the soil quality and results in the death of predators. This in turn harms the food chain thus causing loss of equilibrium in nature. The growth of beneficial insects is encouraged by growing suitable plants which attract them. Beneficial insects are actually predators which control harmful insects. Disease resistant varieties are chosen for cultivation, in order to keep diseases at bay without having to spend money on costly pesticides. Special types of crops known as 'companion crops' are grown to control pests. These crops help in diverting or discouraging the growth of harmful pests. Biological pesticides such as neem extract, are useful in controlling many different pests. The practice of crop rotation helps in disturbing the reproduction cycles of pests, thereby inhibiting their growth and protecting the crops.

(b) Soil quality – An indicator of sustainability

Soil quality has been regarded as a comprehensive index of sustainability to indicate both fertility and eco-system parameters. Of all the physical factors, pH appears to be a very important factor affected by the use of fertilizers.

When nitrogenous fertilizers such as ammonium sulphate, ammonium chloride, diammonium phosphate and urea are added to the soil ammonium ions are released which would react with oxygen to form free hydrogen ions in presence of nitrifying bacteria.

The continuous use of such fertilizers in the soils with poor lime status leads to development of acid soils. The plants can only thrive well if the soil pH is 5.6 to 6.6.

(c) Biodynamics

This method of farming has quite different applications. This method includes the supply of blood meal, bone meal, decomposed horns and nails of animals etc., to the plants. Bone meal is best suited for acidic soils. It contains 25-30% phosphorous and 2.5% lime.

(d) Vermiculture

Soil fauna play a prominent role in regulating soil processes and among these the termites and the earthworms play a vital role in maintaining soil quality and efficient nutrient cycling.

(e) Earthworm: the natural Bioreactor

Earthworms play a key role in soil biology as versatile natural bioreactors. They efficiently harness the beneficial soil micro flora, destroy soil pathogens and convert organic wastes into valuable products such as biofertilizers, biopesticides, vitamins, enzymes, antibiotics, growth hormones and proteinous worm biomass.

(f) Vermicompost

Vermicompost consists of earthworm castings. It is a natural biofertilizer, which consists of 1.5%Nitrogen, 2%Phosphorous and 0.5%Potassium.

A wide range of classification of Biofertilizers is given on next page.

(g) Cultivation

Polyculture is an important aspect of organic farming. In the traditional form of farming, monoculture is practiced, which includes growing a single crop in a given piece of land. Though the motive behind cultivating a single crop, is to reduce costs incurred in fertilizers, seeds, pesticides, etc. However, it creates problems in the long run. The reduction in the fertility of the soil owing to the

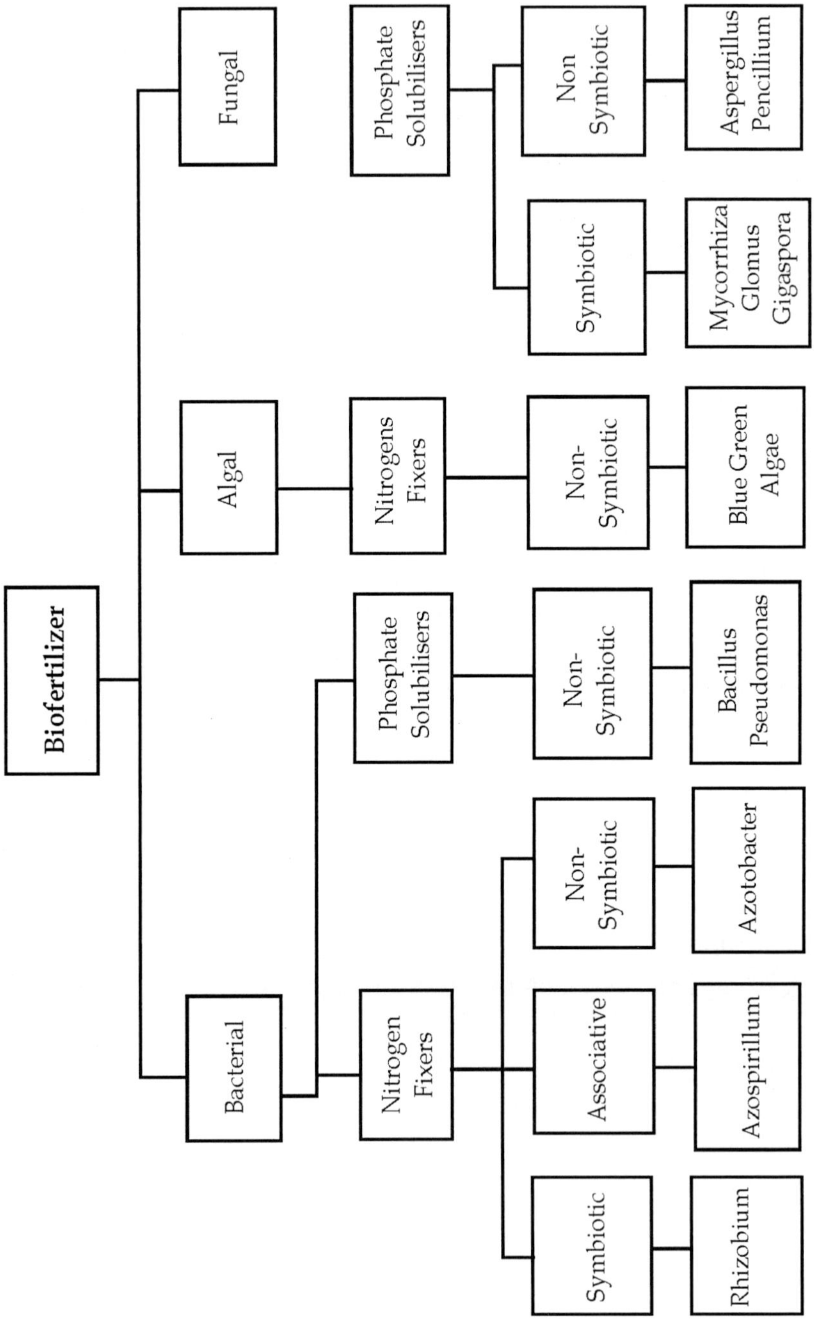
Biofertilizer
Bacterial
Algal
Fungal
Nitrogen Fixers
Phosphate Solubilisers
Nitrogens Fixers
Phosphate Solubilisers
Symbiotic
Associative
Non-Symbiotic
Non-Symbiotic
Non-Symbiotic
Symbiotic
Non Symbiotic
Rhizobium
Azospirillum
Azotobacter
Bacillus Pseudomonas
Blue Green Algae
Mycorrhiza Glomus Gigaspora
Aspergillus Pencillium

extraction of nutrients over a long period and soil erosion result from the practice of monoculture. Moreover, the pests become immune to the chemicals used for their control. Polyculture is a completely different approach towards farming as compared to monoculture. In this method of farming, a variety of crops are cultivated on a single piece of land. It helps attract different soil microbes. Some crops act as repellents to pest and these results in pest control, in an organic manner.

(h) Fertilization

Composts are used to enhance soil fertility in organic farming methods. Green manuring too, is a nice way to add nutrients to the soil. It is the practice of growing plants with prolific leaf growth like Alfalfa, and burying them in the soil before the cultivation of the main crop. The green manuring crops add organic matter to the soil that is necessary for plant growth.

BENEFITS OF ORGANIC FARMING

- Organic farming proves to be more profitable than the age old traditional farming methods.
- It has been found that organic farming reduces the production cost by about 25-30%, as it does not involve the use of synthetic fertilizers and pesticides, which thus makes organic farming cost effective.
- Soil is the most important component in farming and organic farming preserves soil by reducing soil erosion up to a large extent.
- Organic farming also enables the farmers to use the soil for a longer period of time to grow crops as soil fertility is maintained for a long time.
- Organic farming has a positive effect on the ecosystem, as it proves vital in supporting the survival of wildlife in the lowlands. It even provides safe pasture lands for grazing
- Organic farming is not only beneficial for farmers, but it also has proved useful for the dairy industry. Cattle grazing on organic farmlands have been found to be less prone to diseases and they yield more milk. These are definitely good

signs for a consumer of these dairy products from health perspective and for a dairy organization from the profit perspective.

- Products or foodstuffs produced from organic farming neither contain any sort of artificial flavors or preservatives nor do they contain any harmful chemicals.
- The original nutritional content of food is preserved due to the absence of synthetic fertilizers and pesticides.
- Organic products moreover are tastier than the products yielded from traditional farming.
- Consumption of products obtained from organic farming minimizes the risks of physical ailments such as heart attacks. Scientific studies have proven that organic foods are healthier than the inorganic ones.

Organic farming has proven beneficial for many but the yield of organic farming has not been substantial since its beginning. Many farmers have turned to organic farming only on account of the subsidies given by the government to promote organic farming. There also have been instances wherein farmers have opted for organic farming on account of reduced production cost as compared to traditional farming. Low productivity, higher time required to yield and requirement of specialized skills for implementation have been some of the disadvantages of organic farming. However, organic farming has been significantly contributing towards providing quality food and prevention of soil erosion which have nowadays become serious concerns in many regions of the world.

THE FEATURES OF ORGANIC FARMING

It is a common belief that our health is directly related to the food we consume. The quality of the food depends on the health of the soil it grows. Here are a few of the features of organic farming. Organic farming gives importance to environmental protection and helps to sustain ecological issues such as soil conservation. Farmers who undertake organic farming practice crop rotation to enrich the soil with natural mineral resources. Organic farmers have to follow the norms set by the local organic farming associations and they are not allowed to cultivate genetically modified (GM) crops. The minerals for the crop known as crop nutrients are given using insoluble nutrient sources through soil microorganisms that increase nitrogen levels in the soil. For instance, alternating legumes with the main crop would increase nitrogen levels in the soil. Chemical drugs are not administered on farm animals to control fleas or parasite problems. Instead, these problems are controlled by moving the animals to new pastures and by using home remedies to control the plant and animal pests. Organic gardening, including vegetable gardening, is also a part of organic farming. Many flower and vegetable gardens are using composite manure for their flowering plants and shrubs, instead of chemical fertilizers.

REASON FOR CHOOSING ORGANIC PRODUCTS

Today, more and more people are opting for organic food products and organic clothes made from organic cotton and wool. The reasons why organic products are preferred over inorganic products are listed below.

- The organic food products including vegetables, coffee and fruits taste better than the ones produced using chemicals.
- Organic farmers use herbal pesticides, instead of chemical sprays. Hence, the organic fruits and vegetables are safe to eat for the people who suffer from asthma and heart disease.
- Organic farming is environment-friendly and does not use genetically modified hybrids or chemical fertilizers.
- The organic farmers are concerned about the welfare of the farm animals. The milk from an organically fed cow is better than the tinned milk or powdered milk.
- Organic food is more fresh compared to non-organic food because no preservatives are used.

ORGANIC FARMING USING BIOTECHNOLOGY

Now you can apply biotechnology to upgrade the natural system on organic farming. By using biotechnology, farming management is easier than depending on the natural system. It is more effectively applied on the mass production with the same system and result, while costing less. Its organic farming product will be free of chemicals and pesticides.

Biological fertilizer with bacteria composition is the result of biotechnology research and is a good solution for organic farming system. A cell of bacteria is coated by a coat until you use it in farming. On the soil it grows fast by dividing the cells and moving them into the plants roots and build symbiosis system between soil, root, bacteria and fungi.

The composition of biological fertilizer consists of nitrogen binding bacteria, cellulose decomposing bacteria, solvent phosphate bacteria and some specific species of soil bacteria. Those are the best bacteria in the root symbiosis, giving incredible results on organic farming.

The benefits from the biological fertilizer using biotechnology are:

- Bacteria can tie up the nitrogen from the air.
- Optimize the root symbiotic system.
- Ability to suppress the growth of adverse bacteria.
- Produce growth hormone which is useful for plants and supporting plant metabolism system.
- Ability to reduce systemic pesticide.
- Capable to recondition the soil.
- Many species of bacteria have positive ability in affecting the plant by making the leafs become more lush and green.
- Plants growth optimally and increase 30 % plant productivity.
- Free from pesticide or chemical residues product.
- Cut the chemical fertilizer use by more than 40%.
- Cut the harvest time by 1 or 2 weeks.

You can take the benefit and more profits by using biological fertilizer in your organic farming, building a good organic farming system, rehabilitating the soil condition, being environmentally friendly and saving the Earth.

MICROBIAL BIOTECHNOLOGY FOR ORGANIC FARMING

Reasons of health and sustainability of natural/organic farming makes the environment as one alternative to modern agriculture. Organic farming relies on natural ingredients and avoids any intake of synthetic smells, either in the form of synthetic fertilizers, herbicides, or synthetic pesticides. However, farmers often complain of organic agriculture production which tends to lower productivity and more susceptible to pest attack. This problem could be overcome by using microbial-based biotechnology derived from sources of biological wealth of non-synthetic.

Soil is a habitat that is rich in diversity of microorganisms such as bacteria, actinomycetes, fungi, protozoa, algae and viruses. Land-rich farmland contains more than 100 million microbes per gram of soil. Productivity and land carrying capacity depends on the activity of these microbes. Most soil microbes have a profitable role for agriculture. Including soil microbes play a role in degrading organic waste, agricultural waste, re-plant nutrient cycling, biological fixation of nitrogen from the air, dissolving phosphate, stimulating plant growth, plant pathogens biocontrol, helps the absorption of plant nutrients, and establish profitable symbiosis.

USE OF BIOFERTILIZERS AS ORGANIC FERTILIZERS

Organic fertilizers

Microbes play a vital role in facilitating uptake of nutrients in a crop. Rhizobia, among several of these microbes are quite popular for leguminous crops. Microbial inoculation involves the selection and multiplication of plant beneficial micro-organisms and applying them to plants, seed or soil. The main use of micro-organisms is as biofertilizers for improved plant nutrition and as biological control agents to combat pests, weeds and diseases. Microbial inoculation of plants is of great importance in less intensive low-input agricultural systems in developing countries (Davison, 1988).

Biological nitrogen fixation

Biological nitrogen fixation holds much promise for developing countries. Nitrogen fixation is found only in prokaryotic organisms, bacteria and blue-green algae. Rhizobia among several of these microbes are quite popular for leguminous crops. Inoculants for soyabean account for a very large proportion of biological nitrogen fixation. Inoculants for forage, pasture and food legumes are also produced in developing countries. Inoculants use has its main impact on large scale producers (Eaglesham, 1989). Rhizobium inoculation is considered a key component of in the improvement of pasture legume production in South America, West Asia and North Africa and it may prove beneficial elsewhere (Eaglesham, 1989). Pseudomonas and several other bacteria facilitate nutrient uptake and form a very friendly network around the roots. Symbiotic organisms form a harmonious environment in the soils, more abundantly in the tropical climates. It empowers the local community to use its own resources in terms of biodiversity resources and rights, saves on application of high cost fertilizers, renders the soil health and avoids interference with the environment. Biologically active compounds may be liberated from blue-green algae growing on the surface of moist soils. Such compounds may also be released exudates from algae grown in liquid culture. The blue-green algae (Cyanobacteria) are another source of biological nitrogen. They are distributed in symbiosis with the water fern Azolla. Cyanobacteria are structurally diverse assemblages of aerobic gram-negative eubacteria (Prokaryotes) characterized by their ability to form oxygenic photosynthesis. They reduce molecular atmospheric nitrogen to ammonium which can then be utilized for amino acid and protein biosyntheis. Nitrogen compounds which can be assimilated by other plants are ultimately released to the soil after death of the algal cells and subsequent mineralization and nitrification. Algal nitrogen fixation release the fixed nitrogen and its assimilation by higher plants may be rapid (Stewart, 1967). The algal extracts of *Westiellopsis prolifica* analysed by Fogg and Pattnaik (1966) showed that ammonium and amide-nitrogen accounted for most of the total extracellular combined nitrogen. Pretreatment of seeds of vegetable crop with extract of *Westiellopsis prolifica* promoted germination and the subsequent growth of development.

Mycorrhizal associations

Mycorrhizal associations (symbioses between certain fungi and the roots of vascular plants) can increase the rate of uptake of nutrients such as nitrogen and phosphorous from deficient soils. The production of mycorrhizal inocula may well be possible.

Seaweed biofertilizer

Seaweeds are one of the most important marine resources of the world. Seaweed extracts have been marketed for several years as fertilizer additives and beneficial results from their use have been reported (Booth, 1965). The possibilities of using seaweed in modern agriculture has been investigated by many (Thivy, 1961; Aitken and Sen 1965; Boney, 1965). Different forms of seaweed preparation such as LSF (Liquid Seaweed Fertilizers), SLF(Seaweed Liquid Fertilizers), LF (Liquid Fertilizers), and either whole or finally chopped powered algal manure have been used and all of them have been reported to produce beneficial effects on cereals, pulses and flowering plants.

Seaweed manures have the advantage of being free from weeds and pathogenic fungi. Liquid extracts of brown algae are being sold as biostimulants or biofertilizers in various brand names. Promising increased crop yield, nutrient uptake, resistance to frost and stress, improved seed germination of reduced incidence of fungal and inspect attack have been resulted by application of SLF. Seaweeds are known to contain appreciable quantities of plant growth regulators (Mooney and Van Staden, 1985), Cytokinin (Smith and Vanstaden, 1984), IAA (Abe, H. et al., 1972), gibberllins and gibberllin like substance (Bentley, 1960; Radley, 1961; Sekar et al., 1995). Hence marine algae, particularly seaweeds have a vital role to play in agriculture, especially in the third world country where irrational use of chemical fertilizer and pesticides is a cause of concern. Extensive regional tribals would need to be conducted with the product to determine the environmental limits on biological activity and monitor the survival and dispersal of the inocula (Davison, 1988). Hence use of modern agriculture in conjunction with traditional farming practices is the sustainable solution for the future.

GENETICALLY MODIFIED CROPS

The Genetically Modified Crops are the foods that have a gene extracted from a living thing and placed in a different food by a scientist or an expert. The main aim of Genetically Modified Crops is to create a food that is able to survive even if any harmful chemicals or pesticides or herbicides are sprayed. Another purpose of genetically modified crops is to make food stay fresh for a long time.

Gene technology is a type of modern biotechnology that makes use of living things to make or change product. Some of genetically modified crops and food are corn, tomato, beets, potatoes, sprouts, alfalfa, soyabeans. canola oil, rice, wheat, meat, poultry, Processed Foods like Convenience Foods, Condiments, Salad Dressings, Juice, Soda, Vitamins, Chocolate including many frozen meals, heat and serve meals.

The first commercially grown genetically modified whole food crop was a tomato (called FlavrSavr), which was modified to ripen without softening, by Calgene, later a subsidiary of Monsanto. Calgene took the initiative to obtain FDA approval for its release in 1994 without any special labeling, although legally no such approval was required. It was welcomed by consumers who purchased the fruit at a substantial premium over the price of regular tomatoes. However, production problems and competition from a conventionally bred, longer shelf-life variety prevented the product from becoming profitable. A variant of the Flavr Savr was used by Zeneca to produce tomato paste which was sold in Europe during the summer of 1996. The labeling and pricing were designed as a marketing experiment, which proved, at the time, that European consumers would accept genetically engineered foods. Currently, there are a number of food species in which a genetically modified version exists.

TABLE 1

Food	Properties of the genetically modified variety	Modification	Percent Modified in US	Percent Modified in world
Soybeans	Resistant to glyphosate or glufosinate herbicides	Herbicide resistant gene taken from bacteria inserted into soybean	93%	77%
Corn, field	Resistant to glyphosate or glufosinate herbicides. Insect resistance via producing Bt proteins, some previously used as pesticides in organic crop production. Vitamin-enriched corn derived from South African white corn variety M37W has bright orange kernels, with 169x increase in beta carotene, 6x the vitamin C and 2x folate.	New genes, some from the bacterium Bacillus thuringiensis, added/transferred into plant genome.	86%	26%

Contd...

Contd...

Food	Properties of the genetically modified variety	Modification	Percent Modified in US	Percent Modified in world
Cotton (cotton-seed oil)	Pest-resistant cotton	Bt crystal protein gene added/transferred into plant genome	93%	49%
Alfalfa	Resistant to glyphosate or glufosinate herbicides	New genes added/transferred into plant genome.	Planted in the US from 2005-2007; no longer planted currently due to court decisions	
Hawaiian papaya	Variety is resistant to the papaya ringspot virus.	New gene added/transferred into plant genome	80%	
Tomatoes	Variety in which the production of the enzyme polygalacturonase (PG) is suppressed, retarding fruit softening after harvesting.	A reverse copy (an antisense gene) of the gene responsible for the production of PG enzyme added into plant genome	Taken off the market due to commercial failure.	None
Potatoes	*Amflora* variety produces waxy potato starch composed almost exclusively of the amylopectin component of starch.	The gene for granule bound starch synthase (GBSS) (the key enzyme for the synthesis of amylose) was switched off by inserting antisense copy of the GBSS gene.	*Amflora* will be produced solely under contract farming conditions and not made available on the general market.	

Contd...

Contd...

Food	Properties of the genetically modified variety	Modification	Percent Modified in US	Percent Modified in world
Rapeseed (Canola)	Resistance to herbicides (glyphosate or glufosinate), high laurate canola	New genes added/transferred into plant genome	93%	
Sugar cane	Resistance to certain pesticides, high sucrose content.	New genes added/transferred into plant genome		21%
Sugar beet	Resistance to glyphosate, glufosinate herbicides	New genes added/transferred into plant genome	95%	9%
Rice	Genetically modified to contain high amounts of Vitamin A (beta-carotene)	"Golden rice" Three new genes implanted: two from daffodils and the third from a bacterium		

In addition, various genetically engineered micro-organisms are routinely used as sources of enzymes for the manufacture of a variety of processed foods. These include alpha-amylase from bacteria, which converts starch to simple sugars, chymosin from bacteria or fungi that clots milk protein for cheese making, and pectinesterase from fungi which improves fruit juice clarity.

TRANSGENIC STRATEGIES FOR PLANT PROTECTION

Characteristics

The transgenic breeding method or the DNA recombinant technology allows the transference of genes from one species to another without sexual hybridization. Plant breeding by biotechnological processes (Brasileiro & Carneiro 1998) starts with prospecting desirable traits expressed under control of a single gene, e.g., production of a protein that could be toxic to herbivorous insects. Microorganisms and plants that produce toxins against pests have been the main source of genes for this technology. The next step in the process is the identification of the gene in charge of expressing the selected trait and then its isolation using restriction enzymes. This gene is then cloned and linked to other gene fragments to form the cassette of expression. This cassette is a piece of DNA that will be inserted into the plant to promote the expression of the desired characteristic.

Besides the gene of interest the cassette of expression is commonly comprised of a marker gene, a promoter gene and a terminator gene. The promoter gene has a central role in the gene transfer. Its function is to promote the expression of the gene of interest since many foreign genes (mainly those isolated from prokaryotes) will not express in plants. The 35S promoter gene isolated from the cauliflower mosaic virus is largely used due to its strength in inducing the expression of selected genes, and because it has a constitutive character, i.e., it is expressed in all tissues of the plant. Another important gene usually present in the cassette of expression is the marker gene. It allows the selection of the transformed cells in the cell culture media. The most commonly used marker is the antibiotic resistance gene. In this case, the selection is made by adding to the culture media a specific antibiotic. The transformed cells that became resistant to that antibiotic will survive and the other cells will die. Biosafety concerns about the use of antibiotic resistant genes has lead researchers and the industry to avoid these genes and use new markers such as herbicide resistance and the expression of special carbohydrate (mannose) degradation enzymes. The terminator gene contained in the cassette "informs" to the cells the end of the expression code.

The last step in the transformation process is to grow the selected cells to regenerate the plant that will express and transmit the new trait to other varieties through sexual hybridization or cloning.

Insecticidal toxins from *B. thuringiensis* (*Bt*)

During sporulation *B. thuringiensis* produces crystalline inclusions that are protoxins called d-endotoxins. These structures called cry proteins, when ingested by the insect, are dissolved in the gut and cleft by digestive proteases activating the toxin. The toxin then binds to specific glycoprotein receptors on the surface of cells lining the gut, causing an imbalance in ion concentration, destroying the cells, and resulting in the death of the insect (Choma *et al.* 1990). In addition, some strains of *B. thuringiensis* produce "vegetative insecticidal proteins" (VIPs) before sporulation (Lacey & Kaya, 2000).

Since 1942, when E. A. Steinhaus demonstrated the potential of this bacterium for pest control, the use of *Bt* formulations increased as bioinsecticides because of its safety, been harmless to vertebrates and other non-target fauna, and because of the specificity of each cry protein which infects only a narrow range of hosts (Tanada & Kaya 1993, Peferoen 1997). However, instability and degradation of cry proteins when exposed to ultra violet radiation and short persistence on the plant (easily washable by rain and irrigation), have constrained large adoption of *Bt* pesticides by farmers.

B. thuringienis produces a broad diversity of cry proteins with a range of around 100 holotype toxins distributed in 40 groups (cry1, cry2, etc.) each one with several subgroups and a narrow range of host (Bravo *et al.* 1998). This variability is being used to develop transgenic plants resistant to pest attack.

Transgenic plants expressing *Bt* toxins

Several *Bt* toxins have been inserted into crop plants to provide protection against different groups of pest insects. A brief list was presented by Jouanin *et al.* (1998), which we updated (Table 2). Many crops such as vegetables, forage crops, root crops, cereals, and trees are now being transformed to be protected against insects by *Bt* toxins (Shelton *et al.* 2002). New constructs of *Bt* toxin genes with promoters to be expressed in monocots or dicots, and in different tissues of the plant are been studied, including the integration of a native *Bt* gene into the chloroplast genome of tobacco (McBride *et al.* 1995). The chloroplast genome is bacterial in nature, avoiding the need of modifying the toxin gene for higher expression. Although it is still not commercially available, this technique opens new perspectives for the *Bt* plant breeding in the future.

Inhibitors of insect digestive enzymes

Plants extensively produce proteins with antimetabolic activity against diverse digestive enzymes of herbivores. These proteins have either a role as part of defense strategy against herbivores or for preserving stored nutrients as proteins and carbohydrates. Genes encoding these proteins are a valuable source to enhance resistance of plants against insects and could be used in plant breeding programs.

Inhibitors of digestive proteases

There are four classes of protease inhibitors based on their specificity: serine, cysteine, metallo- and aspartyl-proteases. All are small proteins that bind to the insect's digestive enzymes producing an inactive complex that prevents the absorption of amino acids, leading to nutrient starvation. Consequently, these inhibitors induce death or reduction in larval growth. The protease inhibitors can also induce the over-expression of digestive enzymes affecting larval growth. Serine protein inhibitors have activity sites, which inhibit trypsin and chimotrypsin, and are effective antimetabolites against Lepidoptera and Diptera (Duan *et al.* 1986, Hilder *et al.* 1995, Xu *et al.* 1996, Gatehouse *et al.* 1997). A successful case is the cysteine protease inhibitor gene introduced into poplar to protect the plant against beetles (Leplé *et al.* 1995).

Inhibitor of Alfa-Amylase

The alfa-amylase inhibitor is a seed protein contained in the common bean, *Phaseolus vulgaris* L., and its linkage with certain amylases forming inactive complex provide seed protection against several bruchid beetles (Shade *et al.* 1994, Shroeder *et al.* 1995). However, other species of bruchids (*Zabrotes subfasciatus* Boh. and *Acanthoscelides obtectus* Say) have a protective mechanism to inactivate alpha-amylase (Ishimoto *et al.* 1996). The limited range of target insects makes it difficult to evaluate the real value of this gene in the protection of seeds against post-harvest pests.

Lectins

This is a heterogeneous group of proteins with carbohydrate-binding activity. They are found in many plants, in higher concentrations in storage tissues and seeds. Lectin expression genes have been inserted into several plant species which then demonstrated enhanced resistance against insects, although the way these proteins cause deleterious effects on insects is not well known (Gatehouse & Gatehouse 2000). The well-documented anti-nutrient action and toxicity of these proteins towards mammals and birds in nature, in spite of its inactivation by cooking (Shatters 2000), have served as a baseline of an intense discussion about the biosafety of transgenic plants.

Other genes and traits

Transgenic expression of other proteins is also an alternative to the use of *Bt* genes. Chitinase against the beetle *Orizaephilus mercator* Fauvel (Jouanin *et al.* 1998), and cholesterol oxidase against the cotton boll weevil *Anthonomus grandis* Boh., are some examples of enzymes expressed by transgenes with potential effects on insect, although protective effects on plants have still not been demonstrated (Gatehouse & Gatehouse 2000).

TABLE 2

Insect Resistant Transgenic Plants Expressing *B. thuringienis* Toxins

Crop plant	Toxin	Target insect	Reference
Alfaalfa	cry1Ca	*Spodoptera litoralis* (Boiusduval) (Lep.)	Sthrizhov *et al.* (1996)
Broccoli	cry1C	*Plutella xylostella* (L.) (Lep.)	Zhao *et al.* (2001)
Cabbage	cry1Ab	*P. xylostella*	Bhattacharya *et al.* (2002)
Canola	cry1Ac	*Thrichoplusia ni* (Hübner) (Lep.), *Spodoptera exigua* (Hübner), *Heliothis virescens* (Fabr.), *Helicoverpa zea* (Boddie) (Lep.)	Stewart *et al. (1996b)*
	cry1Ac	*P. xylostella*	Ramachandran *et. al* (1998)
Cotton	cry1Ab	*H. virescens, H. zea*	Perlak *et al.* (1990)
	cry1Ac and cry2Ab	*S. exigua, Pseudoplusia includens* (Walker (Lep.)	Adamezyk *et al.* (2001)
Eggplant	cry1Ab	*Leucinodes orbonalis* Guenée (Lep.)	Kumar *et al.* (1998)
	cry3A	*Leptinotarsa decemlineata* (Say) (Col.)	Jelenkovic *et al.* (1998)
Corn	cry1Ab	*Ostrinia nubilalis* (Hübner) (Lep.)	Koziel *et al.* (1993)
	cry9c	*O. nubilalis*	Jansem *et al.* (1997)
Poplar	cry1Aa	*Lymantria dispar* (L.) (Lep.)	McCown *et al.* (1991)
	cry3Aa	*Chrysomela tremulae* F. (Col.)	Cornu *et al.* (1996)

Contd...

Contd...

Crop plant	Toxin	Target insect	Reference
Potato	cry1Ab	*Phthorimaea operculella* (Zeller) (Lep.)	Peferoen *et al.* (1992), Rico *et al.* (1998)
	cry1Ab	*Heliothis armigera* (Hübner)	Chakrabarti *et al.* (2000)
	cry3Aa	*L. decemlineata*	Adang *et al.* (1993), Perlak *et al.* (1993), Coombs *et al.* (2002)
Rice	cry1Ab	*Chilo supressalis* Walker (Lep.)	Fujimoto *et al.* (2000)
	cry1B	*C. supressalis* *Cnaphalocrosis medinalis* Guenée (Lep.)	Marfa *et al.* (2002) Wunn *et al.* (1996)
	cry1Ac, cry2A and	*C. medinalis, Scirpophaga incertulas* Walker (Hom.)	
	GNA	*Nilaparvata lugens* Stal (Hom.)	Maqbool *et al.* (2001)
	cry1Ab and cry1Ac	*C. supressalis*	Chang *et al.* (1998)
Soybean	cry1Ac	*H. virescens, H. zea* *P. includens*	Stewart *et al.* (1996a)
Tobacco	cry1Aa	*Manduca sexta* (L.) (Lep.)	Barton *et al.* (1987)
	cry1Ab	*M. sexta*	Vaeck *et al.* (1987)
	cry1Ab and cpT1	*M. sexta*	Perlak *et al.* (1991)
	cry1Ab	*M. sexta*	Williams *et al.* (1993)
	cry1Ac	*H. virescens, H. zea,* *S. litoralis*	McBride *et al.* (1995)
	cry1C	*S. litoralis*	Strizhov *et al.* (1996)
	cry2A	*H. armigera*	Selvapandiyan *et al.* (1998)
	cry2A	*H. virescens, H. zea* *S. exigua*	Kota *et al.* (1999)
Tomato	cry1Ab	*H. virescens*	Fischolff *et al.* (1987)
	cry1Ac	*H. armigera*	Mandaokar *et al.* (2000)

Lep = Lepidoptera; Col = Coleoptera; Hom = Homoptera

GM PRODUCTS : BENEFITS AND CONTROVERSIES

Benefits

Crops

- Enhanced taste and quality
- Reduced maturation time
- Increased nutrients, yields, and stress tolerance
- Improved resistance to disease, pests, and herbicides
- New products and growing techniques

Animals

- Increased resistance, productivity, hardiness, and feed efficiency
- Better yields of meat, eggs, and milk
- Improved animal health and diagnostic methods

Environment

- "Friendly" bioherbicides and bioinsecticides
- Conservation of soil, water, and energy
- Bioprocessing for forestry products
- Better natural waste management
- More efficient processing

Society

- Increased food security for growing populations

Controversies

Safety

- Potential human health impacts, including allergens, transfer of antibiotic resistance markers, unknown effects
- Potential environmental impacts, including: unintended transfer of transgenes through cross-pollination, unknown effects on other organisms (e.g., soil microbes), and loss of flora and fauna biodiversity

Access and Intellectual Property

- Domination of world food production by a few companies
- Increasing dependence on industrialized nations by developing countries
- Biopiracy, or foreign exploitation of natural resources

Ethics

- Violation of natural organisms' intrinsic values
- Tampering with nature by mixing genes among species
- Objections to consuming animal genes in plants and vice versa
- Stress for animal

Labeling

- Not mandatory in some countries (e.g., United States)
- Mixing GM crops with non-GM products confounds labeling attempts

Society

- New advances may be skewed to interests of rich countries

ORGANIC FARMING AND CHEMICAL FARMING : A COMPARITIVE STUDY

- Chemical farming deals with the use of chemicals in agriculture. This is the modern technique of farming. On the contrary Organic Farming deals with the use of natural products and discard of chemicals. Organic farming is believed to be about 10,000 years old whereas chemical farming came into being after World War I.
- Chemical farming has deteriorating effects on the environment while Organic farming is Eco-friendly.
- Though chemical farming is nutrient specific it is hazardous to health while organic farming does not pose any side effects.
- The farmer spraying the chemicals gets affected while organic farming does not involve the use of chemicals.
- Chemical farming is not economical, as it demands huge expenditure on chemicals. Organic techniques involve the use of plant debris and other organic wastes and thus prove to be economical.
- The organic farming practices use sustainable methods i.e. they do not harm natural quality of the environment. On the contrary chemical farming intervenes in the natural harmony.
- Organic pest control methods are non hazardous whereas chemical pest control methods have far reaching effects.

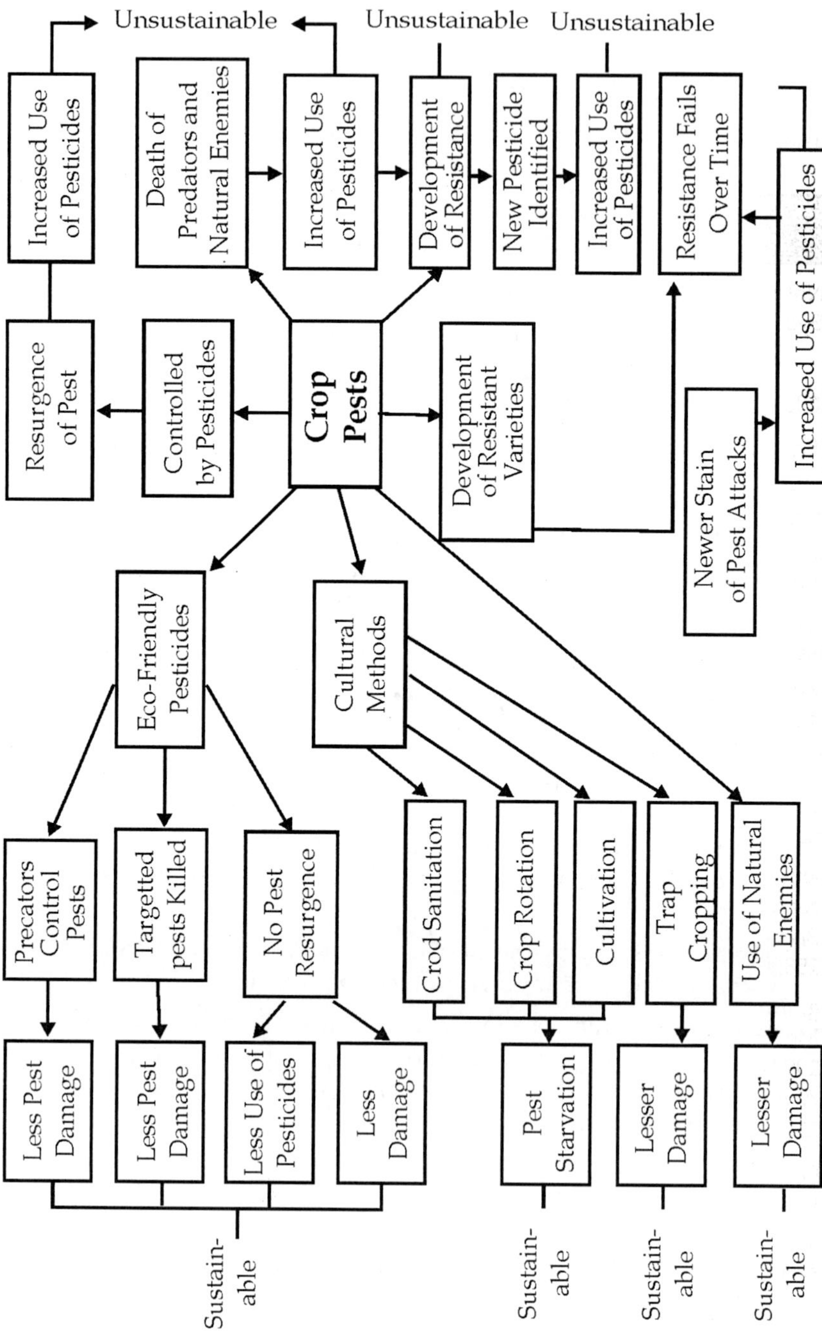
Crop Pests
Controlled by Pesticides
Resurgence of Pest
Increased Use of Pesticides
Unsustainable
Death of Predators and Natural Enemies
Increased Use of Pesticides
Unsustainable
Development of Resistance
Unsustainable
New Pesticide Identified
Increased Use of Pesticides
Unsustainable
Development of Resistant Varieties
Resistance Fails Over Time
Newer Stain of Pest Attacks
Increased Use of Pesticides
Eco-Friendly Pesticides
Precators Control Pests
Targetted pests Killed
No Pest Resurgence
Less Pest Damage
Less Pest Damage
Less Use of Pesticides
Less Damage
Sustain-able
Cultural Methods
Crod Sanitation
Crop Rotation
Cultivation
Trap Cropping
Use of Natural Enemies
Pest Starvation
Sustain-able
Lesser Damage
Sustain-able
Lesser Damage
Sustain-able

- Chemicals used in chemical farming are eco unfriendly. But the organic wastes used in organic farming decompose at a faster rate and thus are eco friendly.

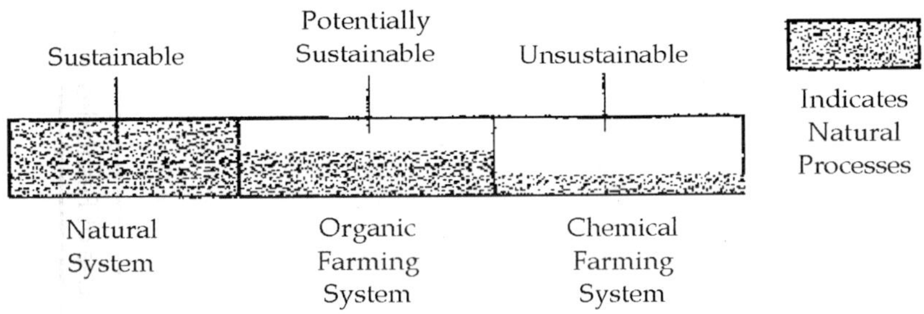

CONCLUSION

The harmony within nature we are seeking now is central to the ethos of the pre industrial age. The message to the restless youth of today and tomorrow to be happy and healthy is to move towards a conservation society. The most difficult but the most essential form of structural adjustment we need today is adjustment to sustainable life styles. Sustainable agronomy is impossible without farmers and against nature. Everything in the natural world belongs to the Nature Goddess. We owe no rights towards them but owe a sense of responsibility towards the love of Mother Nature.

REFERENCES

Abe, H., Vchiyams, M. and Sato, R., 1972. Isolation and identification of nature action in marine algae. Agro. Biol. Chem. 36:2259-2260.

Adamczyk Jr., J.J., L.C. Adams & D.D. Hardee. 2001. Field efficacy and seasonal expression profiles for terminal leaves of single and double *Bacillus thuringiensis* toxin cotton genotypes. J. Econ. Entomol. 94: 1589-1593.

Adang, M.J., M.S. Brody, G. Cardineau, L.J. Petersen, G.B. Parker, S.A. McPherson, J. Wyman, S. Love, G. Reed, D. Biever & D.A. Fischholf. 1993. The recontruction and expression of *Bacillus thuringiensis* cry III A gene in protoplants and potato plants. Plant Mol. Biol. 21: 1131-1145.

Aitken, J.E. and Senn, J.L. 1965. Seaweed products as fertilizer and sal conditions. Bot. Mar. 8:144- 148.

Alstad, D.N. & Andow, D.A. 1995. Managing the evolution of insect resistance to transgenic plants. Science 268: 1894-1896.

Armer, C.A., Berry, R.E. and Kogan, M. 2000. Longevity of phytophagous heteropteran predators feeding on transgenic Bt-potato plants. Entomol. Exp. Appl. 95: 329-333.

Arpaia, S. 1996. Ecological impact of Bt-transgenic plants: 1. Assessing possible effects of cryIIIB toxin on honey bee (*Apis mellifera* L.) colonies. J. Genet. Breed. 50: 315-319.

Atkins, E.L., Kellum, D. and Atkins, K.W. 1981. Reducing pesticide hazards to honey bees: mortality prediction techniques and integrated management strategies. Leaflet 2883, revised edition. Riverside, Division of Agriculture Sciences, University of California,24p.

B. Nanda, Tripathy, S.K. and Padhi, S. 1991. Effect of algalization on seed germination of vegetable crops. World Journal of microbiology and biotechnology 7:622-623.

Baker H.G. 1974. The evolution of weeds. Annu. Rev. Ecol. Syst. 5: 1-24.

Barton, K.A., Whiteley, R.H. and Yang, N.S. 1987. *Bacillus thuringiensis* d-entodoxin expressed in transgenic *Nicotiana tabacum* provides resistance to lepidopteran insects. Plant Physiol. 85: 1103-1109.

Bell, H.A., Fitches, E.C., Down, R.E., Ford, L., Marris, G.C., Edwards, J.P., Gatehouse, J.A. & Gatehouse, A.M.R. 2001. Effect of dietary cowpea trypsin inhibitor (CpTI) on the growth and development of the tomato moth Lacanobia oleracea (Lepidoptera: Noctuidae) and on the success of the gregarious ectoparasitoid *Eulophus pennicornis* (Hymenoptera: Eulophidae). Pest Manag. Sci. 57: 57-65.

Bentley, J.A. 1960. Plant hormones in marine phytoplankton, zooplankton and seawater. J. Mar. Biol. Ass. U.K., 39:433-444.

Bhattacharya, R.C., Viswakarma, N., Bhat, S.R., Kirti, P.B. & Chopra, V.L. 2002. Development of insect-resistant transgenic cabbage plants expressing a synthetic cryIA(b) gene from Bacillus thuringiensis. Curr. Sci. 83: 146-150.

Birch, A.N.E., Geoghegan, I.E., Majerus, M.E.N., McNicol, J.W., Hackett, C. Gatehouse, A.M.R. and Gatehouse, J.A. 1999. Tri-trophic interactions involving pest aphids, predatory 2-spot ladybirds and transgenic potatoes expressing snowdrop lectin for aphid resistance. Mol. Breed. 5: 75-83.

Boney, A.D. 1965. A biology of marine algae. Hutchinson on Educational Ltd., London.

Booth, E. 1965. The manorial value of seaweed. Bot. Mar. 8:138-143.

Bourget, D., Chaufaux, J., Micoud, A., Delos, M., Naibo, B., Bombarde, F., Marque, G., Eychenne, N. and Pagliari, C. 2002. *Ostrinia nubilalis* parasitism and the field abundance of non-target insects in transgenic *Bacillus thuringiensis* corn (*Zea mays*). Environ. Biosafety Res. 1: 49-60.

Brasileiro, A.C.M. & Carneiro, V.T. de. 1998. Manual de transformação genética de plantas. Embrapa SPI/Cenargen, Brasília, 309p.

Bravo, A., Sarabia, S., Lopez, L., Ontiveros, H., Abarca, C., Ortiz, A., Ortiz, M., Lina, L., Villalobos, F.J., Peña, G., Nuñes-Valdez, M.E., Soberon, M. and Quintero, R. 1998. Characterization of cry genes in a mexican Bacillus thuringiensis strain collection. Appl. Environ. Microb. 64: 4965-4972.

Cannon, R.J.C. 2000. Bt transgenic crops: risks and benefits. Int. Pest. Manag. Rev. 5: 151-173.

Carpenter, J., Felsot, A., Goode, T., Hammig, M., Onstad, D. and Sankula, S. 2002. Comparative environmental impacts of biotechnology-derived and traditional soybean, corn, and cotton crops. Council for Agricultural Science and Technology. Iowa, USA. 189p.

Chakrabarti, S.K., Mandaokar, A.D., Shukla, A., Pattanayak, D., Naik, P.S., Sharma, R.P. and Kumar, P.A. 2000. *Bacillus thuringiensis* cry1Ab gene confers resistance to potato against *Helicoverpa armigera* (Hubner). Potato Res. 43: 143-152.

Cheng, X.Y., Sardana, R., Kaplan, H. and Altosaar, I. 1998. Agrobacterium-transformed rice plants expressing synthetic cryIA(b) and cryIA(c) genes are highly toxic to striped stem borer and yellow stem borer. Proc. Natl. Acad. Sci. USA 95: 2767-2772

Choma, C.T., Surewicz, W.K., Carey, P.R., Pozsgay, M., Raynos, T. and Kaplan, H. 1990. Unusual proteolysis of the protoxin and toxin of *Bacillus thuringiensis* - structural implications. Eur. J. Biochem. 189: 523-527.

Christensen, M. 1989. A view of fungal ecology. Mycologia 81: 1-19.

Coombs, J.J., Douches, D.S., Li, W.B., Grafius, E.J. and Pett. W.L. 2002. Combining engineered (Bt-cry3A) and natural resistance mechanisms in potato for control of Colorado potato beetle. J. Am. Soc. Hort. Sci. 127: 62-68.

Cornu, D., Leplé, J.C., Bonadé-Bottino, M., Ross, A., Augustin, S., Delplanque, A., Jouanin, L. and Pilate, G. 1996. Expression of proteinase inhibitor and a Bacillus thuringiensis d -endotoxin in transgenic poplars, p.131-136. In Proceedings IUFRO Meeting on Somatic Cell Genetics and Molecular Genetic of Trees. Kluwer, Dordrecht, 287p.

Cui, J. and Xia, J. 2000. Effects of Bt (*Bacillus thuringiensis*) transgenic cotton on the dynamics of pest population and their enemies. Acta Phytoph. Sin. 27: 145-52.

Dale, P.J., Clarke, B. and Fontes, E.M.G. 2002. Potential for the environmental impact of transgenic crops. Nat. Biotechol. 20: 567-574.

Davison, J. 1988. Plant beneficial bacteria. Biotechnology, 6 March, 1988, 282-286.

Dogan, E.B., Berry, R.E., Reed, G.L. and Rossignol, P.A. 1996. Biological parameters of convergent lady beetle (Coleoptera: Coccinellidae) feeding on aphids (Homoptera: Aphididae) on transgenic potato. J. Econ. Entomol. 89: 105-1108.

Duan, X., Li, X., Xue, Q., Abo-El-Saad, M., Xu, D. and Wu, R. 1996. Transgenic rice plants harboring an introduced potato proteinase inhibitor II gene are insect resistant. Nat. Biotechnol. 14: 494-496.

Ellstrand, N.C., Prentice, H.C. and Hancock, J. 1999. Gene flow and introgression from domesticated plants into their wild relatives. Annu. Rev. Ecol. Sys. 30: 539-563.

ESA. 2001. Entomological Society of America. Symposium Summaries. 2001 ESA Annual Meeting. December 9-12, San Diego, California.

Fischholff, D.A., Bowdish, K.S. Perlak, F.J. Marrone, P.G. McCormick, S.M., Niedermeyer, J.G., Dean, D.A., Kusano-Katzmer, K., Mayer, E.J., Rochester, D.E., Rogers, S.G. and Finley, R.T. 1987. Insect tolerant transgenic tomato plants. BioTechnology 5: 807-813.

Fitter, A., Perrins, J. and Williamson, M. 1990. Weed probability challenged. BioTechnology 8: 473.

Fogg, G.E. and Pattnaik, H. 1966. The release of extracellular nitrogenous products by *Westiellopsis prolifica* janet. Phykos, 5:58-67.

Fujimoto, H., Itoh, K., Yamamoto, M., Kyozuka, J. and Shimamoto, K. 1993. Insect resistant rice generated by introduction of a modified o- endotoxin gene of *Bacillus thuringiensis*. BioTechnology 11: 1151–1155.

Gatehouse, A.M.R., Davison, G.M., Newell, C.A., Merryweather, A., Hamilton, W.D.O., Burgess, E.P.J., Gilbert, R.J.C. and Gatehouse, J.A. 1997. Transgenic potato plants with enhanced resistance to the tomato moth, Lacanobia oleracea, growthroom trials. Mol. Breed. 3: 49-63.

Gatehouse, J.A. and Gatehouse, A.M.R. 2000. Genetic engineering of plants for insect resistance, p. 211-241. In J.E. Rechcigl & N.A. Rechcigl (eds.). Biological and biotechnological control of insect pests. Lewis, Boca Raton, p. 374.

Giles, K.L., Hellmich, R.L., Iverson, C.T. and Lewis, L.C. 2000. Effects of transgenic Bacillus thuringiensis corn grain on *B. thuringiensis*-susceptible *Plodia interpunctella* (Lepidoptera: Pyralidae). J. Econ. Entomol. 93: 1011-1016.

Girard, C., Picard-Nizou, A.L., Grallien, E., Zaccomer, B., Jouanin, L. and Pham-Delegue, M.H. 1998. Effects of proteinase inhibitor ingestion on survival, learning abilities and digestive proteinases of the honeybee. Transgenic Res. 7: 239-246.

Gould, F. 1991. The evolutionary potential of crop pests. Am. Sci. 79: 496-507.

Gould, F. 1998. Sustainability of transgenic insecticidal cultivars: integrating pest genetics and ecology. Annu. Rev. Entomol. 43: 701-726.

Greenpeace Report. 2002. A summary of research on the environmental impact of Bt cotton in China. D. Xue (ed.). Published by Greenpeace, 26p.

Groot, A.T. and Dicke, M. 2002. Insect-resistant transgenic plants in a multi-trophic context. Plant J. 31: 387-406.

Hails, R.S., Reeds, M., Kohn, D.D. and Crawley, M.J. 1997. Burial and seed survival in *Brassica napus* subsp. *Oleifera* and *Sinapis arvensis* including a comparison of transgenic and non-transgenic lines of the crop. Proc. R. Soc. Lond. Ser. B, Biol. Sci. 264: 1-7.

Head, G., Surber, J.B., Watson, J.A., Martin, J.W. and Duan, J.J. 2002. No detection of cry1Ac protein in soil after multiple years of transgenic Bt cotton (Bollgard) use. Environ. Entomol. 31: 30-36.

Hilbeck A, Moar, W.J., Pusztai-Carey, M., Filippini, A.M. and Bigler, F. 1998b. Toxicity of *Bacillus thuringiensis* cry1Ab toxin to the predator Chrysoperla carnae. Environ. Entomol. 27:1255-63.

Hilbeck, A. 2002. Transgenic host plant resistance and non-target effects, p. 167-185. In D.K. Letourneau & B.E. Burrows (eds.). Genetically engineered organisms. Assessing environmental and human health effects. CRC Press, Boca Raton, FL, 438p.

Hilbeck, A., Baumgartner, M., Fried, P.M. and Bigler, F. 1998a. Effects of transgenic Bacillus thuringiensis corn-fed prey on mortality and development time of immature *Chrysoperla carnea* (Neuroptera: Chrysopidae). Environ. Entomol. 27: 480-487.

Hilbeck, A., Moar, W.J., Pusztai-Carey, M., Filippini, A. and Bigler, F. 1999. Prey-mediated effects of cry1Ab toxin and protoxin and cry2A protoxin on the predator *Chrysoperla carnea*. Entomol. Exp. Appl. 91: 305-316.

Hilder, V.A., Powell, K.S., Gatehouse, A.M.R., Gatehouse, J.A., Gatehouse, L.N., Shi, Y., Hamilton, W.D.O., Merryweather, A., Newell, C., Timans, J.C., Peumans, W.J., Van Damme, E.J.M. and Boulter, B. 1995. Expression of snowdrop lectin in transgenic tobacco plants results in added protection against aphids. Transgenic Res. 4: 18-25.

Hoy, C.W., Feldman, J., Gould, F., Kennedy, G.G., Reed, G. and Wyman, J.A. 1998. Naturally occurring biological controls in genetically engineered crops, p. 185-205. In P. Barbosa (ed.). Conservation biological control. Academic Press, London, 396p.

Ishimoto, M., Sato, J., Chrispeels, M.J. and Kitamura, K. 1996. Bruchid resistance of transgenic azuki bean expressing seed alpha-amylase inhibitor in the common bean. Entomol. Exp. Appl. 79: 309-315.

James, C.A. 2001. Global review of commercialized transgenic crops, 2001. ISAAA Briefs No. 24. Preview ISAAA, Ithaca, NY.

James, R.R., Miller, J.C. and Lighthart, B. 1993. *Bacillus thuringiensis* var. kurstaki affects a beneficial insect, the cinnabar moth (Lepidoptera: Archtiidae). J. Econ. Entomol. 86: 334-339

Jansen, S., vanVliet, A., Dickburt, C., Buysse, L., Piens, C., Saey, B., de Wulf, A., Gossele, V., Paez, A. and Gobel, E. 1997. Transgenic corn expressing a cry9C insecticidal protein from *Bacillus thuringiensis* protected from European corn borer damage. Crop Sci. 37: 1616-1624.

Jelenkovic, G, Billings, S., Chen, Q., Lashomb, J., Hamilton, G. and Ghidiu, G. 1998. Transformation of eggplant with synthetic cryIIIA gene produces a high level of resistance to the Colorado potato beetle. J. Am. Soc. Hort. Sci. 123: 19-25.

Johnson, K.S., Scriber, J.M., Nitao, J.K. and Smitley, D.R. 1995. Toxicity of *Bacillus thuringiensis* var. kurstaki to three non-target Lepidoptera in field studies. Environ. Entomol. 24: 288-297.

Jouanin L., Bonade-Bottino, M., Girard, C. and Morrot, G. 1998. Transgenic plants for insect resistance. Plant Sci. 131: 1-11.

Kogan, M. 1986. Plant defense strategies and host-plant resistance, p. 83-134. In M. Kogan (ed.), Ecological theory and integrated pest management. John Wiley & Sons, New York, 362p.

Kota, M., Daniell, H., Varma, S., Garczynski, S.F., Gould, F. and Moar, W.J. 1999. Overexpression of the *Bacillus thuringiensis* (Bt) cry2Aa2 protein in chloroplasts confers resistance to plants against susceptible and Bt-resistant insects. Proc. Natl. Acad. Sci. USA 96: 1840-1845.

Koziel, M.G., Beland, G.L, Bowman, C., Carozzi, N.B., Crenshaw, R., Crosslaind, L., Dawson, J., Desai, N., Hill, M., Kadwell, M., Launis, K., Lewis, K., Maddox, D., Mc Pherson, D., Meghiji, M.R., Merlin, E., Rhodes, R., Warren, G.W., Wright,. M. and Evola, S.V. 1993. Field perfomance of elite transgenic corn plants expressing insecticidal protein derived from Bacillus thuringiensis. BioTechnology 11: 194-200.

Kumar, P.A., Mandaokar, A., Sreenivasu, K., Chakrabarti, S.K., Bisaria, S., Sharma, S.R., Kaur, S. and Sharma, R.P. 1998. Insect-resistant transgenic brinjal plants. Mol. Breed. 4: 33-37.

Lacey, L.L. and Kaya, H.K. 2000. Field manual of techniques in invertebrate pathology. Dordrecht, Kluwer, 911p.

Leplé J.C., Bonadé-Bostino, M., Augustin, S., Pilate, G., Le Tân, V.D., Delplanque, A., Cornu, D. and Jouanin, L. 1995. Toxicity to *Chrysomela tremulae* (Coleoptera: Chrysomelidae) of a transgenic poplars expressing a cysteine proteinase inhibitor. Mol. Breed. 1: 319-328.

Lewis, W.J. and Takasu, K. 1990. Use of learned odours by a parasitic wasp in accordance with host and food needs. Nature 348: 635-636.

Losey, J. E., Rayor, L.S. and Carter, M.E. 1999. Transgenic pollen harms monarch larvae. Nature 399: 214.

Lozzia, G.C. 1999. Biodiversity and structure of ground beetle assemblages (Coleoptera Carabidae) in Bt corn and its effects on non target insects. Boll. Zool. Agraria Bachiocoltura 31: 37-58.

Lozzia, G.C., Furlanis, C., Manachini, B. and Rigamonti, I.E. 1998. Effects of Bt corn on *Rhopalosiphum padi* L. (*Rhynchota Aphididae*) and its predator *Chrysoperla carnea* Stephen (Neuroptera Chrysopidae). Boll. Zool. Agraria Bachicoltura 30: 153-164.

Luby, J.J. and McNichol, R.F. 1995. Gene flow from cultivated to wild raspberries in Scotland: developing a basis for risk assessment for testing and deployment of transgenic cultivars. Theor. Appl. Genet. 90: 1133-1137.

Malone, L.A., Burgess, E.P.J., Gatehouse, H.S., Voisey, C.R., Tregidga, E.L. and Philip, B.A. 2001. Effects of ingestion of a *Bacillus thuringiensis* toxin and a trypsin inhibitor on honeybee fligth activity and longevity. Apidologie 32: 57-68.

Mandaokar, A.D., Goyal, R.K., Shukla, A., Bisaria, S., Bhalla, R., Reddy, V.S., Chaurasia, A., Sharma, R.P., Altosaar, I. and Kumar, P.A. 2000. Transgenic tomato plants resistant to fruit borer (Helicoverpa armigera Hübner). Crop Prot. 19: 307-312.

Maqbool, S.B., Riazuddin, S., Loc, N.T., Gatehouse, A.M.R., Gatehouse, J.A. and Christou, P. 2001. Expression of multiple insecticidal genes confers broad resistance against a range of different rice pests. Mol. Breed. 7: 85-93.

Marfa, V., Mele, E., Gabarra, R., Vassal, J.M., Guiderdoni, E. and Messeguer, J. 2002. Influence of the developmental stage of transgenic rice plants (cv. Senia) expressing the cry1B gene on the level of protection against the striped stem borer (*Chilo suppressalis*). Plant Cell Rep. 20: 1167-1172.

Marvier, M. 2001. Ecology of transgenic crops: Genetically engineered plants might generate weed problems and affect non-target organisms, but measuring the risk is difficult. Am. Scien. 89: 160-167.

McBride, K.E., Svab, Z., Schaaf, D.J., Hoogan, P.S., Stalker, D.M. and Maliga, P. 1995. Amplification of a chimeric *Bacillus* gene in chloroplasts leads to an extraordinary level of an insecticidal preotein in tobacco. BioTechnology 13: 362-365.

McCown, B.H., McCabe, D.E., Russll, D.R., Robison, D.J., Barto, K.A. and Rafffa, K.F. 1991. Stable tranformation of Populus and incorporation of pest resistance by electric discharge particle acceleration. Plant Cell Rep. 9 : 590–594.

Miller, J.C., 1990. Field assessment of the effects of a microbial pest control agent on non-target Lepidoptera. Am. Entomol. 36: 1355-139.

Mooney, P.A. and Van Staden, J. 1985. Effect of seaweed concentrate on the growth of wheat under condition of water fern. S. Afr. J. Sci. 8:632-633.

Munkvold, G.P., Hellmich, R.L. and Rice, L.G. 1999. Comparison of fumonisin concentration in kernels of transgenic Bt corn hybrids and nontransgenic hybrids. Plant Dis. 83: 130-138.

NRC - National Research Council. 2000. Genetically Modified Pest-Protected Plants: Science and Regulation. National Academy Press. Washington, DC, 263p.

Obrycki, J.J., Losey, J.E., Taylor, O.R. and Jesse, L.C.H. 2001. Transgenic insecticidal corn: beyond insecticidal toxicity to ecological complexity. BioScience 51: 353-361.

Orr, D.B. and Landis. D.A. 1997. Oviposition of European corn borer (Lepidoptera: Pyralidae) and impact of natural enemy populations in transgenic versus isogenic corn. J. Econ. Entomol. 90: 905-909.

Peferoen, M. 1992. Engineering of insect-resistant plants with Bacillus thuringiniesis crystal protein genes. Biotech. Agric. 7: 135-153.

Peferoen, M. 1997. Insect control with transgenic plants expressing *Bacillus thuringiensis* crystal proteins, p. 21-48. In N. Carrozi & M.G. Coziel (eds.), Advances in insect control: the role of transgenic plants. Taylor and Francis, London, 301p.

Perlak, F.J., Fuchs, R.L., Dean, D.A., McPherson, S.L. and Fishholff, D.A. 1991. Modification of the coding sequence enhances plant expression of insect control protein genes. Proc. Natl. Acad. Sci. USA 88: 3324–3328.

Perlak, F.J., Deaton, R.W., Armstrong, T.O., Fuchs, R.L., Sims, S.R., Greenplate, J.T. and Fishholff, D.A. 1990. Insect resistant cotton plants. BioTechnology 8: 939-943.

Perlak, F.J., Stone, T.B., Muskopf, Y.M., Eagan, N., Roush, R.T. Shewmaker, C.K., Jones, A., Oakes, J.V. and McBride, K.E. 1993. Geneticaly improved potatoes: protection from damage by Colorado potato beetles. Plant Mol. Biol. 22: 313–321.

Picard-Nizou, A.L., Grison, R. Olsen, L., Arnold, C.P.G. and Pham-Delegue, M.H. 1997. Impact of proteins used in plant genetic engineering: toxicity and behavioral study in the honeybee. J. Econ. Entomol. 90: 1710-1716.

Pilcher, C.D., Obrycki, J.J., Rice, M.E. and Lewis, L.C. 1997. Preimaginal development, survival, and field abundance of insect predators on transgenic *Bacillus thuringiensis* corn. Environ. Entomol. 26: 445-454.

Pimentel, D.S. and Raven, P.H. 2000. Bt corn pollen impacts on non-target Lepidoptera: assessment of effects in nature. Proc. Natl. Acad. Sci. USA 97: 8198-8199.

Price, P. W. 1997. Insect ecology. John Wiley & Sons, New York, NY, 874p.

Price, P.W., Bouton, C.E., Gross, P., McPherson, B.A., Thompson, J.N., and Weis, A.E. 1980. Interactions among three trophic levels: interactions between insect herbivores and natural enemies. Annu. Rev. Ecol. Syst. 11: 41-65.

Radley, M. 1961. Gibberllin-like substances in plants. Nature, London 191:684-685.

Ramachandran, S., Buntin, G.D., All, J.N., Raymer, P.L. and Stewart, C.N. 1998. Greenhouse and field evaluations of transgenic canola against diamondback moth, *Plutella xylostella*, and corn earworm, Helicoverpa zea. Entomol. Exp. Appl. 88: 17-24.

Ramachandran, S., Buntin, G.D., All, J.N., Raymer, P.L. and Stewart, C.N. 2000. Integrated pest management: Intraspecific competition of an insect-resistant transgenic canola in seed mixtures. Agron. J. 92: 368-374.

Raymond, M., Callaghan, A., Fort, P. and Pasteur, N. 1991. Worldwide migration of amplidied insecticide resistance genes in mosquitoes. Nature 350: 151–153.

Regal, P.J. 1988. The adaptative potential of genetically engineered organisms in nature. Trends Ecol. Evol. 3: S36-S38.

Regal. P.J. 1994. The true meaning of "exotic species" as a model for genetically engineered organisms. Experientia 49: 225-229.

Rice, M.E. and Pilcher, C.D. 1998. Potential benefits and limitations of transgenic Bt corn for management of the European corn borer (Lepidoptera: Crambidae). Am. Entomol. 44: 74-78.

Rico, E., Ballester, V. and Mensua, J.L. 1998. Survival of two strains of *Phthorimaea operculella* (Lepidoptera: Gelechiidae) reared on transgenic potatoes expressing a Bacillus thuringiensis crystal protein. Agronomie 18: 151-155.

Riddick, E.W. and Barbosa, P. 2000. Cry3A-intoxicated *Leptinotarsa decemlineata* (Say) are palatable prey for Lebia grandis Hentz. J. Entomol. Sci. 35: 342-346.

Riddick, E.W., Diveli, G. and Barbosa, P. 2000. Season-long abundance of generalist predators in transgenic vs. non-transgenic potato fields. J. Entomol. Sci. 35: 349-359.

Saxena, D. and Stotzky, G. 2000. Insecticidal toxin from *Bacillus thuringiensis* is released from roots of transgenic Bt corn in vitro and in situ. FEMS Microbiol. Ecol. 33: 35-39.

Saxena, D., Flores, S. and Stotzky, G. 1999. Insecticidal toxin in root exudates from Bt corn. Nature 402: 480.

Schuler, T.H. 2000. The impact of insect resistant GM crops on populations of natural enemies. Antenna 24: 59-65.

Schuler, T.H., Poppy, G.M. and Denholm,. I. 2000. Recommendations for assessing effects of GM crops on non-target organisms, p.1221-1228.

BCPC Conference - Pests and Diseases 2000, Harpenden, UK, IACR-Rothamsted.

Schuler, T.H., Poppy, G.M., Potting, R.P.J., Denholm, I. and Kerry, B.R. 1999a. Interactions between insect tolerant genetically modified plants and natural enemies. BCPC Symp. Proc. 72: 197-202.

Schuler, T.H., Potting, R.P.J., Denholm, I. and Poppy, G.M. 1999b. Parasitoid behaviour and Bt plants. Nature 400: 825-826.

Sears, M.K., Hellmich, R.L., Stanley-Horn, E., Oberhauser, K., Pleasants, J.M., Mattila, H.R., Siegfried, B.D. and Dively, G.P. 2001. Impact of Bt corn pollen on monarch butterfly populations: A risk assessment. Proc. Natl. Acad. Sci. USA 98: 11937-11942.

Sekar, R. Thangaraju, N. and Rangasamy, R. 1995. Effect of seaweed liquid fertilizer from *Ulva lactuca* L. on *vigna unguiculata* L. (Walp) Phykos. 34:49-53.

Selvapandiyan, A., Reddy, V.S., Kumar, P.A., Tewari, K.K. and Bhatnagar, R.K. 1998. Transformation of Nicotiana tabacum with a native cry1Ia5 gene confers complete protection against *Heliothis armigera* Mol. Breed. 4: 473-478.

Shade, R.E., Schroeder, H.E., Pueyo, J.J., Tabe, L.M., Murdock, L.L., Higgins, T.J.V. and Chrispeels, M.J. 1994. Transgenic pea seeds expressing the alpha-amylase inhibitor of the common bean are resistant to bruchid beetles. Biotechnology 12: 793-796.

Shatters, Jr., R.G. 2000. Environmental impact of biotechnology, p. 281-302. In J.E. Rechcigl & N.A. Rechcigl (eds.), Biological and biotechnological control of insect pests. Lewis Publishers, Boca Raton, FL, USA, 374p.

Shelton, A.M., Zhao, J.Z. and Roush, R.T. 2002. Economic, ecological, food safety, and social consequences of the deployment of Bt transgenic plants. Annu. Rev. Entomol. 47: 845-881.

Shroeder, H.E., Gollasch, S., Moore, A., Tabe, L.M., Craig, S., Hardie, D.C., Chrispeels, M.J., Spencer, D. and Higgins, T.J.V. 1995. Bean alpha-amylase inhibitor confers resistance to pea weevil (*Bruchus pisorum*) in transgenic peas (*Pisum sativum* L.). Plant Physiol. 107: 1233-1239.

Smith, F.B.C. and Van staden, J. 1984. The effect of seaweed concentrate and fertilizer on growth and endogenous Cytokinin content of *Phaseolus vulgaris*. South African journal of Botany, 3:375-379.

Snow, A.A. 2002. Transgenic crops – why gene flow matters. Nature Biotech. 20: 452.

Stewart Jr., C.N., All, J.N., Raymer, P.L. and Ramachandran, S. 1997. Increased fitness of transgenic insecticidal rapeseed under insect selection pressure. Mol. Ecol. 6: 773-779.

Stewart, C.N., Adang, M.J., All, J.N., Raymer, P.L., Ramachandran, S. and

Parrott, W.A. 1996b. Insect control and dosage effects in transgenic canola containing a synthetic *Bacillus thuringiensis* cry IAc gene. Plant Physiol. 112: 115-120.

Stewart, W.D.P. 1967. Transfer of biologically fixed nitrogen in a sand dune slack region. Nature 214, 603-604.

Stewart. C.N., Adang, M.J., All, J.N., Boerma, H.R., Cardineaux, G., Tucker D. and Parrot, W.A. 1996a. Genetic tranformation recovery, and characterization of fertile soybean transgenic for a synthetic *Bacillus thuringiensis* cry IA (c) gene. Plant Physiol. 112: 121-129.

Stotzky, G. 2000. Persistence and biological activity in soil of insecticidal proteins from *Bacillus thuringiensis* and of bacterial DNA bound on clays and humic acids. J. Environ. Qual. 29: 691-705.

Stotzky, G. 2002. Release, persistence, and biological activity in soil of insecticidal proteins from Bacillus thuringiensis, p. 187-222. In D.K. Letourneau & B.E. Burrows (eds.), Genetically engineered organisms. Assessing environmental and human health effects. CRP Press, Boca Raton, FL, 438p.

Strizhov, N., Keller, M., Mathur, J., Koncz-Kalman, K., Bosch, D., Prudovsky, E., Schell, J., Sneh, B., Koncz, C. and Zilberstein, A. 1996. A synthetic cry IC gene, encoding a *Bacillus thuringiensis* d-endotoxin, confers Spotoptera resistance in alfalfa and tobacco, Proc. Nalt.Acad Sci. USA 93: 1512-15017.

Tabashnik, B.E. 1994. Evolution of resistance to *Bacillus thuringiensis*. Annu. Rev. Entomol. 39: 47-79.

Tanada, Y. and Kaya, H.K. 1993. Insect pathology. Academic Press, San Diego, 666p.

Thivy, F. 1961. Seaweed manure for perfect sal and smiling fields. Salt. Res. Indust. 1:14.

Tiedje, J.M., Colweell, R.K., Grossman, Y.L., Hodson, R.E., Lenski, R.E., Mack, R.N. and Regal, P.J. 1989. The planned introduction of genetically engineered organisms: Ecological considerations and recommendations. Ecology 70: 298-315.

US EPA (US Environmental Protection Agency). 2000. Sets of scientific issues being considered by the Environmental Protection Agency regarding Bt plant-pesticides risk and benefit assessments. SAP Report. 2000-07. FIFRA Scientific Advisory Panel Meeting, October 18-20, 2000.

Vaeck, M., Reynaerts, A., Hofte, H., Jansens, S., DeBeuckleer, M., Dean, C., Zabeau, M., Van Montagu, M. and Leemans, J. 1987. Transgenic plants protected from insect attack. Nature 327: 33-37.

Vinson, S.B. 1976. Host selection by insect parasitoids. Ann. Rev. Entomol. 21: 109-134.

Whaley, W.H., Anhold, J. and Schaalje, G.B. 1998. Canyon drift and dispersion of *Bacillus thuringiensis* and its effects on selected non-target lepidopterans in Utah. Environ. Entomol. 27: 539-548.

Williams, S., Friedrich, L., Dincher, S., Carozzi, N., Kessmann, H., Ward E. and Ryals, J. 1993. Chemical regulation of *Bacillus thuringiensis* delta-endotoxin expression in transgenic plants. BioTechnology 7: 194-200.

Williamson, M. 1993. Invaders, weeds and the risk from genetically manipulated organisms. Experientia 49: 219-224.

Williamson, M., Perrings, J. and Fitter, A. 1990. Releasing genetically engineered plants: present proposals and possible hazards. Trends Ecol. Evol. 5: 417-419.

Wraight, C.L., Zangerl, A.R., Carroll, M.J. and Berenbaum, M.R. 2000. Absence of toxicity of *Bacillus thuringiensis* pollen to black swallowtails under field conditions. Proc. Natl. Acad. Sci. USA 97: 7700-7703.

Wunn, J., Kloti, A., Burkhardt, PK., Biswas, G.C.C., Launis, K., Inglesias V.A. and Potrykus, I. 1996. Transgenic indica rice breeding line IR 58 expressing a synthetic cry A (b) gene from *Bacillus thuringiensis* provides effective pest control. BioTechonology 14: 171- 176.

Xu, D.P., Xue, Q.Z., McElroy, D., Malwal, Y., Hilder, V.A. and Wu, R. 1996. Constitutive expression of a cowpea trypsin-inhibitor gene, CpT1, in transgenic rice plants confers resistance to 2 major rice insect pests. Mol. Breed. 2: 167-173.

Zangerl, A.R., McKenna, D., Wraight, C.L.. Carrol, M., Ficarello, P., Warner, R. and Berenbaum, M.R. 2001. Effects of exposure to event 176 *Bacillus thuringiensis* corn pollen on monarch and black swallowtail caterpillars under field conditions. Proc. Natl. Acad. Sci. USA 98: 11908-11912.

Zhao, J.Z., Li, Y.X., Collins, H.L., Cao, J., Earle, E.D. and Shelton, A.M. 2001. Different cross-resistance patterns in the diamondback moth (Lepidoptera: Plutellidae) resistant to *Bacillus thuringiensis* toxin crylC. J. Econ. Entomol. 94: 1547-1552.

Zwahlen, C., W. Nentwig, F. Bigler & A. Hillbeck. 2000. Tritrophic interactions of transgenic Bacillus thuringiensis corn, Anaphothrips obscurus (Thysanoptera: Thripidae), and the predator *Orius majusculus* (Heteroptera: Anthocoridae). Environ. Entomol. 29: 846-850.

❑❑❑

Index

❑❑❑